Theory and Practice of Thermoelectric Thermometry

Springer
*Singapore
Berlin
Heidelberg
New York
Barcelona
Budapest
Hong Kong
London
Milan
Paris
Tokyo*

Handbook of Temperature Measurement • Volume 3

Theory and Practice of Thermoelectric Thermometry

Robin E. Bentley

Springer

CSIRO

Robin E. Bentley
National Measurement Laboratory
CSIRO Telecommunications and Industrial Physics
P O Box 218
Lindfield
New South Wales 2070
Australia

Library of Congress Cataloging-in-Publication Data

Handbook of temperature measurement / edited by Robin E. Bentley
    p.   cm.
Includes bibliographical references
ISBN 9814021121 (set)

1. Temperature measurements--Handbooks, manuals, etc.
2. Hygrometry--Handbooks, manuals, etc. 3. Thermometers--Handbooks, manuals, etc. 4. Thermocouples -- Handbooks, manuals, etc. I. Bentley, Robin E. II. v. 1. Temperature and humidity measurement   III. v. 2. Resistance and liquid-in-glass thermometry IV. v. 3. The theory and practice of thermoelectric thermometry.
QC271 .H276 1998
536/.5/0287 21   98-28324
                                                                                      CIP

ISBN 981-4021-11-3 (Volume 3)
ISBN 981-4021-12-1 (set)

This work is subject to copyright. All rights are reserved, whether the whole or part of the material is concerned, specifically the rights of translation, reprinting, reuse of illustrations, recitation, broadcasting, reproduction on micro-films or in any other way, and storage in databanks or in any system now known or to be invented. Permission for use must always be obtained from the publisher in writing.

© Springer-Verlag Singapore Pte. Ltd. 1998
Printed in Singapore

The publisher makes no representation, express or implied, with regard to the accuracy of the information contained in this book and cannot accept any legal responsibility or liability for any errors or omissions that may be made.

Typesetting: Camera-ready by author
SPIN 10688216 (Volume 3)   5 4 3 2 1 0
SPIN 10688224 (set)   5 4 3 2 1 0

Preface

Temperature is one of the most widely measured physical quantities. Its accurate measurement is essential for the safe and efficient operation and control of a vast range of industrial processes. This book, one of three volumes in the Handbook of Temperature Measurement, is intended to give the reader an understanding not just of the appropriate techniques and instrumentation, but also of the underlying principles of measurement and sensor design. A discussion of the international framework for measurement traceability and the temperature scale is also given. Despite this, the emphasis is most focussed on the practical aspects of thermometry, such as calibration techniques and the identification and minimisation of error sources.

The books grew out of the biennial "Temperature Measurement Course" run by the Australian National Measurement Laboratory (NML). Started in the 1940's, the 3-5 day courses are given to participants from industry and laboratories involved in metrology and research, from Australia and the Asia-Pacific region. Each of the chapters is written by NML staff lecturing at this course, and expert in their field.

The Australian National Measurement Laboratory is a National Facility within the CSIRO (Commonwealth Scientific and Industrial Research Organisation), Division of Telecommunication and Industrial Physics. Established in 1938, it is located in Sydney and has responsibility for the maintenance and dissemination of the national units of measurement, for overseeing national measurement traceability and the international traceability of Australia's national standards. The temperature standards group within NML has played a leading role in international thermometry research, conducts industrial research projects and takes an active role in assisting industry in industrial measurement problem solving. This close contact with industry is reflected in the practical approach to thermometry given in these books.

Barry D. Inglis
Director
National Measurement Laboratory, Australia
May 1998

Contents

	Preface	v
1	**The Emf in Thermocouples**	**1**
1.1	Introduction	1
1.2	Thermoelectricity	3
	1.2.1 Case 1: the electric circuit	4
	1.2.2 Case 2: thermal conduction	7
1.3	The thermocouple and its emf	11
	1.3.1 Emf distribution	13
	1.3.2 Inhomogeneity	13
	1.3.3 Changes in immersion	16
	1.3.4 Laws of thermoelectric circuits	17
1.4	Standard reference tables for thermocouples	17
	1.4.1 The use of reference tables when $CJ \neq 0°C$	19
	1.4.2 Use of reference tables in diagnosis	20
2	**Thermocouple Materials and their Properties**	**25**
2.1	Introduction	25
2.2	Conventional thermocouple types	25
	2.2.1 Manufacturing tolerances	27
2.3	Properties of conventional thermoelements	30
2.4	Elemental thermocouples	33
2.5	Platinum-based thermocouples (types B, R and S)	35
	2.5.1 Effects in bare-wires up to 1200°C	36
	2.5.2 Effects in bare wires beyond 1200°C	38

		2.5.3 Long-term drift data (bare-wire)	40
		2.5.4 MIMS probes of type B, R or S	41
	2.6	Nickel-based thermocouples (types K and N)	41
		2.6.1 Thermoelectric stability	45
		2.6.2 Bare-wire thermocouples	48
		2.6.3 MIMS probes	52
		2.6.4 MIMS probes sheathed in Inconel or stainless steel	54
		2.6.5 Integrally-designed MIMS probes	56
		2.6.6 Properties of ID-MIMS thermocouples	59
		2.6.7 Availability of ID-MIMS probes	66
		2.6.8 MIMS: type K or type N?	67
	2.7	Thermocouples based on Constantan (types E, J and T)	69
	2.8	Thermocouples for very low and very high temperatures	70
	2.9	Tip formation for bare-wire thermocouples	73
	2.10	Insulation and protection for bare-wire thermocouples	74
		2.10.1 Flexible insulation	75
		2.10.2 Rigid insulation	76
		2.10.3 Protection tubes and sheaths	76
3	**Thermocouples in Use**		**81**
	3.1	Introduction	81
	3.2	Real thermocouples	82
	3.3	The cold junction	83
	3.4	Extension leads	85
		3.4.1 Extension lead polarity	87
		3.4.2 Extension lead calibration	88
	3.5	Measurement of emf	89
	3.6	Switching	92
	3.7	Thermocouples in parallel	96
	3.8	Thermocouples in series	99
	3.9	Differential thermocouples	99
	3.10	The measurement of gas temperature	103

CONTENTS

4 The Calibration of Thermocouples — 113
- 4.1 Introduction — 113
- 4.2 Calibration — 114
  - 4.2.1 Points to consider in thermocouple calibrations — 115
- 4.3 Reference standard thermocouples — 118
  - 4.3.1 Hysteresis state: quenched? — 119
  - 4.3.2 Calibration of the standard — 120
  - 4.3.3 NML calibration procedure — 122
- 4.4 Laboratory calibration — 123
  - 4.4.1 Thermoelectric scanning — 126
- 4.5 A suggested calibration procedure — 129
- 4.6 Interpolation — 132
  - 4.6.1 Software for generating calibration results — 135
- 4.7 Calibration uncertainty — 137
  - 4.7.1 An example — 140
- 4.8 Use of a calibration report — 142
- 4.9 Movable test probes — 143
- 4.10 *In-situ* calibration — 145
- 4.11 The calibration of instruments — 147

5 The Uncertainty in Temperature Measurement — 153
- 5.1 Introduction — 153
- 5.2 Error, accuracy or uncertainty? — 154
- 5.3 Error characterisation — 156
- 5.4 Treatment of uncertainties — 159
  - 5.4.1 The ISO guide — 160
  - 5.4.2 The ISO guide: a summary — 166
  - 5.4.3 The ISO guide: an example — 167
  - 5.4.4 The low-fuss method — 170
- 5.5 Rounding — 173
- 5.6 Measurement: a summary — 174
- 5.7 Sources of systematic error — 175
- 5.8 Experimental assessment of systematic error — 184
- 5.9 Statistical treatments of random error — 186

6 Multi-Site Temperature Measurement — **189**

- 6.1 Introduction 189
- 6.2 Materials specifications 190
- 6.3 Interpretation of temperature specifications 191
- 6.4 Standard AS 2853 195
  - 6.4.1 Development of AS 2853 195
  - 6.4.2 The AS 2853 approach 201
  - 6.4.3 A sample enclosure test 202

A The Thermodynamics of Thermoelectricity — **207**

- A.1 General equations of flow 207
- A.2 The Kelvin relations 208
- A.3 Seebeck effect 209
- A.4 Peltier and Thomson effects 210
- A.5 Absolute Seebeck coefficients 211
- A.6 Some calculations for copper 213
  - A.6.1 The electric case 214
  - A.6.2 The thermal case 215

B Standard Reference Functions for Thermocouple Emf — **219**

- B.1 Introduction 219
- B.2 Reference functions 219
- B.3 Converting V to T 221

Bibliography — **237**

List of Tables

1.1	Values of Peltier and Thomson coefficients	7
1.2	Values of Seebeck coefficient	9
1.3	Sample from thermocouple reference tables	18
2.1	Thermocouple types in common use	26
2.2	Seebeck coefficients for common thermocouples	27
2.3	Manufacturing tolerances	28
2.4	Useful properties of various materials	29
2.5	Seebeck coefficient of Ni-based alloys	32
2.6	Seebeck coefficient of Pt-Rh alloys	33
2.7	Long-term drift in Pt-based thermocouples	40
2.8	Drift in Ni-based thermocouples: bare-wire	51
2.9	Composition of ceramics used for sheathing	78
3.1	Gas temperature error: unshielded probe	106
3.2	Gas temperature error: radiation shield	109
3.3	Probe temperature in ovens: relative to gas and wall	111
3.4	Temperature at which radiation & convection are equal	111
4.1	Uncertainty in using extension leads	125
4.2	Sample calibration data	133
4.3	Calibration results	135
4.4	Calibration data in temperature units	136
4.5	Component uncertainties in a TC calibration	141
4.6	Calibration data for a type K instrument	151
5.1	An uncertainty analysis using the ISO guide	170

5.2	Values of t_p and m_p	188
6.1	Uncertainty limits for enclosure test	200
6.2	Sample enclosure test data	204
A.1	Absolute thermoelectric scale for platinum	214
B.1	Emf difference between 1990 & 1968 reference tables	223
B.2	Maximum error in using rounded coefficients	223
B.3	Standard coefficients for types B & E	224
B.4	Standard coefficients for types J & K	225
B.5	Standard coefficients for types N & R	226
B.6	Standard coefficients for types S & T	227
B.7	Reference tables for type B	228
B.8	Reference tables for type E	229
B.9	Reference tables for type J	230
B.10	Reference tables for type K	231
B.11	Reference tables for type N	232
B.12	Reference tables for type R	233
B.13	Reference tables for type S	234
B.14	Reference tables for type T	235
B.15	Reverse coefficients for type T	236
B.16	Reverse coefficients for type K	236

Chapter 1

The Emf in Thermocouples

1.1 Introduction

The thermocouple, an abbreviation of the earlier term 'thermo-electric couple', is a pair of dissimilar metals joined at one or two places and elsewhere isolated electrically from each other. It produces a small electrical signal that may lead to a good estimate of temperature for one of the points of contact.

The thermocouple arose from experimental demonstrations by Seebeck in 1821 [1]. Seebeck was one of several scientists caught up in a wave of enthusiasm throughout Europe over a discovery announced in 1820: Oersted had demonstrated that an electric current flowing in a wire lying parallel and in close proximity to a magnetic needle will deflect the needle. Within months of this announcement the first electromagnet was proposed and Ampere showed that the electric currents in two conductors had a magnetic influence on each other. Magnetism and electricity were related and some people were convinced that one process was a manifestation of the other.

Seebeck met Oersted and reproduced his experiments. In doing so he noted some peculiarities which, in turn, led him to warm the junction between a copper wire and a bismuth plate, joined in a closed loop. A magnetic needle placed nearby deflected and the deflection reversed on cooling the junction. The deflection also depended on the choice of metals and seemed directly related to the temperature difference between the two junctions. Seebeck referred to the phenomenon as 'thermo-magnetism' and firmly believed that the magnetic effects he had observed were a direct consequence of the temperature difference. He thought there was no electric current involved and took exception to the term 'thermo-electricity' when it was introduced. Oersted described Seebeck's work as "the most beautiful of the discoveries which have yet grown out of mine" [2].

Carrying on from Seebeck, Becquerel studied different combinations of metals heated by a spirit lamp and suggested that temperature could be measured in this way, giving platinum with palladium as the best choice. However, the first 'industrial' thermocouple was possibly the 'magnetic pyrometer' proposed in 1836 [3]. It consisted of a platinum wire sealed into the breech (iron) of a gun, the wire being insulated from it by magnesia or asbestos filling, and the circuit was closed by connecting the two metals via the coil of a simple galvanometer. The breech was inserted into a furnace and the deflection of the galvanometer needle observed. This pyrometer proved unreliable, partly through the use of insensitive galvanometers, and as a consequence the thermocouple method was soundly condemned: a major setback.

The 'thermocouple method' was re-appraised about 50 years later [4] in Europe and independently in America. As a result, platinum with the alloy platinum 10% rhodium was proposed by Le Chatelier for accurate pyrometry and platinum against platinum 20% iridium was suggested by Barus. Le Chatelier made the interesting comment that in constructing 'thermoelectric couples' certain metals such as iron, nickel and palladium should be avoided. They "gave rise to singular anomalies" [4].

An interesting and thorough account of this early history of the thermocouple and of thermoelectricity is given by Finn [2]. He describes the experimental and theoretical struggles of these early pioneers. The lack of adequate theory failed to offer experimental guidance and, just as significantly, the indifferent purity of the materials gave inconsistent data, since all thermoelectric phenomena are highly sensitive to contamination.

Since then hundreds of different thermocouple types have been investigated. Kinzie [5] presents a comprehensive summary of data and literature references for about 300 varieties.

Let us now consider the thermocouple as it is seen at the present time. The user regards it as one of several temperature measuring devices that operate on some electrical property. In most 'electrical thermometers' the temperature sensitive component being measured is localised in a small object, usually referred to as the 'sensor'. These include such things as resistance thermometers, thermistors, integrated-circuit chips of various kinds and the base-to-emitter junctions of transistors. In each case there is a clearly defined sensor connected to two or more leads. The leads are there to convey information from the sensor, which produces the signal, to an appropriate instrument.

Thermocouples present a different picture. The user is aware that it consists of two wires and nothing more, there being no localised sensor in evidence! The wires are usually joined together at one end, the 'tip', and at the other, open end connection is made, often via another pair of leads, to an instrument of high input impedance. The site of the signal is not immediately obvious. Nonetheless, because of our experience with other devices, it is tempting to assume that the tip, formed by joining the two wires together, is in some way responsible.

It is known that the signal is an electro-motive force (emf) and not an electric current—a current may occur, but only as a consequence of closing the circuit. The question of where the emf is produced must be considered. It is important, not just for academic reasons, but for the user of thermocouples as well. Assumptions about the sites of emf within a thermocouple dictate, to a large extent, one's approach to temperature measurement and to the diagnosis of problems. For example, if the tip of a thermocouple is thought to be the site of its emf the user is likely to be anxious about the formation, quality and care of this tip and to be unconcerned about the individual wires. Welding, which produces a mixed alloy at the tip, would be a worry, as would brazing or soldering, which introduces foreign elements. Further, if the thermocouple signal deteriorates, as it is likely to, the cause will be attributed to changes in the tip, and the simple remedy of removing and re-making the tip will not work. Moreover, the alternative and equally 'obvious' solution of sending the thermocouple to a suitable laboratory for calibration, the more costly option, will prove just as unwise (see section 1.4.2).

It is shown in the following pages that emf is not generated at the tip but instead is distributed along the wires. Even among 'experts' and researchers studying thermocouple behaviour, this fact has not always been appreciated [6, 7, 8, 9]. This is partly because the notion of tip-emf is tempting and partly through misconceptions propagated in the early literature [10]. The notion continues in recent publications—even the ASTM in a manual on thermocouples [11] supported a tip-emf until its recent update in 1993.

1.2 Thermoelectricity

In practice there are a variety of electrical signals developed within a thermocouple circuit. All but one are unwanted and unrelated to its functioning as a temperature sensor. These signals, contributing error to the measurement, are dealt with in section 5.7. For temperature measurement we need from the thermocouple its thermoelectric signal, and it is useful to see how the underlying process for this signal fits in with the other processes of thermoelectricity.

Thermoelectricity is that field of study dealing with phenomena in which the processes of heat and electricity overlap and inter-relate. There are three thermoelectric effects, each named after its discoverer. They are the Seebeck, Peltier and Thomson effects. The Seebeck effect is the generation of emf within a conductor whenever heat is flowing, and the Peltier and Thomson effects refer to the production of reversible heat when an electric current flows. Peltier heat appears at the junctions between conductors and Thomson heat appears within each conductor. All three effects are thermodynamically reversible, i.e., if the direction of the cause is reversed so also is the direction of the response. This contrasts, for example, with ohmic heating of electrical components: an irreversible and lossy process that does not reverse (to cooling) on reversing the electric current. Interestingly, all three thermoelectric effects are small in metals and easily missed.

I shall now illustrate and explain these effects with reference to the metal copper. The description is simple, maybe too simple to be strictly valid, but it does serve the purpose. A more formal treatment of thermoelectricity is given in Appendix A, which also contains, in section A.6, a derivation of the values given below.

In a metal it is the electron that causes electrical and thermal conduction. This is not surprising since electrons have both charge and kinetic energy; an electric current is simply the flow of charge and heat flow the transfer of energy.

Electrons behave much like a gas in the sense that they have high individual velocities, collide frequently and yet, as a group (of molecules or electrons), the 'gas' or 'cloud' may appear stationary. When the cloud moves its 'wind' speed will be considerably less than the individual velocities of the component particles. Continuing with this analogy we see that the conductors in an electrical circuit represent corridors along which the electron gas would move when suitably encouraged.

In the absence of external influences, such as electric fields and temperature gradients, the electron cloud is stationary: it has no net motion in any direction. By contrast, those electrons available for conducting electricity or heat, about 0.3% of the valence electrons in copper, have at room temperature an average speed of 6×10^6 km h^{-1} and collide at the rate of 5×10^{13} per second. This chaotic behaviour represents a significant amount of kinetic energy.

1.2.1 Case 1: the electric circuit

Consider a 1 m long copper wire of 1 mm diameter placed in a closed loop and at a uniform temperature of 20°C. If a battery is added to the circuit it produces an electric field along the wire and thus a force on each electron because of its charge. Instantaneously, the electron cloud begins to drift, at 4 cm min^{-1} for a current of 10 A, and 2.2 W of resistive heating is generated in the loop.

A drifting electron cloud necessarily transports its internal kinetic energy

1.2. THERMOELECTRICITY

and thus represents a 'thermal' current that, for this simple example, would continue around the loop indefinitely, trapped within the metal, none of it passing to the surroundings and thus not directly observable. This intrinsic energy flow, which I refer to as the 'Peltier flow', does not contribute to what is normally referred to as heat flow (thermal conduction: page 7) and, for copper, it is 0.006 W for 10 A. For further discussion on the Peltier flow see page 211.

The expected outcome of applying an electric field is an electric current and some resistive heating. The Peltier flow is unexpected, easily missed but just as natural. For an electric current i the Peltier flow is Πi, where Π is the Peltier coefficient of the material and, because the coefficient is both temperature and material dependent, the Peltier flow causes two thermoelectric phenomena. They are the Peltier and the Thomson effects. Some values of Π are given in Table 1.1.

Suppose a length of iron wire is inserted in our circuit and the electric field in the copper portion is maintained. The electric current will have the same value, 10 A, in both metals but the Peltier flows will differ (see Figure 1.1). As the electron cloud passes from iron, where the Peltier flow is 0.039 W, to copper, where it is 0.006 W, the excess, 0.033 W, is given up at the junction, which then increases in temperature. At the other junction the electron cloud must extract heat, so cooling the junction, on entering the iron wire where the Peltier flow is larger. This reversible process at the junctions of dissimilar materials is the **Peltier effect**. It is useful as a heat pump in applications such as air conditioning and refrigeration.

Practical quantities of heat cannot be 'pumped' with metals—their Peltier coefficients are too small and their thermal conductivities too large. For large values of i a large cross section is needed and this results in the rapid dissipation, by conduction, of any heat deposited at a junction. Semiconductors are used for Peltier heat pumps because they have large values of Π and small coefficients of thermal conductivity.

In a conductor having a non-uniform temperature distribution, the Peltier flow, Πi, will also vary continuously along the wire, as the electron cloud progresses, because of the temperature dependence of Π (see Figure 1.1). As it passes into a region of different temperature where, for example, a smaller Π applies, the electron cloud will give up energy to its surroundings. This reversible generation or absorption of heat within a conductor is the **Thomson effect** and is proportional to the electric current and to the temperature gradient. It is also proportional to the Thomson coefficient, some values of which are given in Table 1.1. In a copper wire, 1.8 mW will be produced in each centimetre of its length for an electric current of 10 A and a temperature gradient of $100°C \, cm^{-1}$.

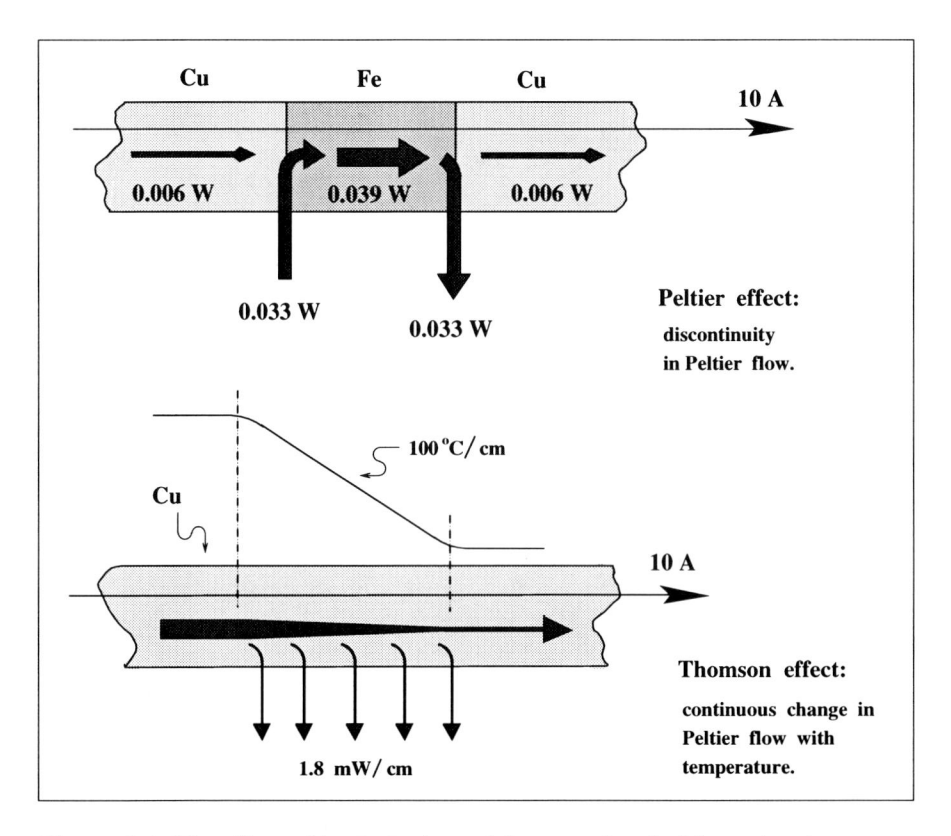

Figure 1.1: The effects of intrinsic thermal flow associated with an electric current.

The Thomson coefficient is a very useful property because it is directly measurable in a single conductor, unlike the other thermoelectric effects that appear as a difference between the contributions from two conductors. From known relationships between the three thermoelectric coefficients measurements of Thomson coefficient, for a particular material, yield absolute values for the other two (see section A.5 on page 211). Some values of Thomson coefficient are given in Table 1.1.

Thermocouples are used as temperature measuring devices under open-circuit conditions, or near to it. Even in moving-coil instruments that require a small electric current, the degree of Peltier and Thomson heating in metallic thermocouples is extremely small and as such has an insignificant effect on temperature measurement. The only thermoelectric effect of relevance to the thermocouple is the Seebeck effect, discussed next.

1.2. THERMOELECTRICITY

Table 1.1: Values of Peltier and Thomson coefficients at 20°C (for their derivation see section A.5).

Conductor	Peltier Coefficient (mV)	Thomson Coefficient (μV K^{-1})
Chromel	6.5	5.9
Fe	3.9	−7.9
Nicrosil	3.5	5.6
Cu	0.6	1.8
Pt	−1.4	−9.1
Nisil	−4.3	−4.7
Alumel	−5.3	−6.7
Ni	−5.7	−16
Constantan	−11	−22

NOTE: for description of above alloys see section 2.2

1.2.2 Case 2: thermal conduction

If a metal bar is heated in such a way that the temperature at one end is higher than that at the other, heat will flow along the bar and the flow is readily observable. This process, known as thermal conduction, is handled mainly by the electrons, first as they become more thermally energetic, as the temperature of the hotter end is raised, and then as they transfer energy to their neighbours.

To illustrate the mechanisms involved, consider the simplistic and unrealistic event of a temperature gradient of $100°\text{C cm}^{-1}$ being instantaneously set up along a copper bar with a cross sectional area of $1\,\text{cm}^2$ (Figure 1.2). Two processes would begin at once. One is a collision process, whereby electrons transfer energy to their less energetic neighbours, and the second is the movement of the electron cloud towards the cooler end, taking its energy with it. The first is the dominant effect, observed as thermal conduction, and the second is small, short-lived and the cause of thermocouple emf. Initially, the drift in the cloud would be extremely small, about $0.3\,\text{cm min}^{-1}$, yet it amounts to a large transfer of charge, equivalent to 110 A, and some Peltier flow. If all the Peltier flow were available for heat transfer to the surroundings it would contribute only 0.015% to the 400 W of heat conducted by the collision process.

As soon as the electron cloud starts to drift it begins accumulating an excess of negative charge at one end of the bar and of positive charge at the other. This causes an electric field acting against further movement of the cloud. The charge reservoirs, and thus the electric field, build up till finally

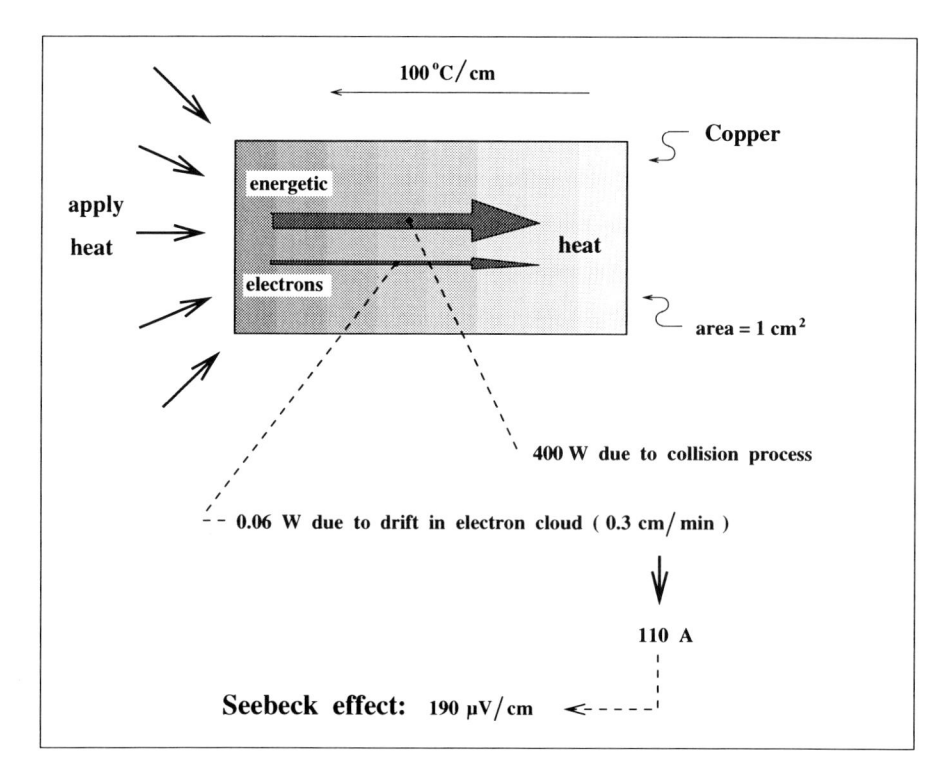

Figure 1.2: One consequence of heat flow is the Seebeck effect: an electric field produced as the electrons endeavour to transport heat.

the electron cloud ceases to drift. The response to the temperature step set up along the bar is now complete and the process is known as the **Seebeck effect**. It represents a balancing act between the tendency for electrons to drift in the 'thermal field', the temperature gradient, and the electric field that results from their initial attempt at movement. The Seebeck effect is a source of power, albeit small in metals, and can instantly accommodate changes in circuit loading. As seen above for the conditions indicated it has the potential to deliver 110 A of electric current. Thus, if a resistive load is placed across the bar the cloud drift instantly adjusts to a suitable value and the electric field is maintained.

The electric field, E, produced by a temperature gradient, dT/dx, is proportional to the gradient:

$$E = S\frac{dT}{dx}. \tag{1.1}$$

This expression defines the Seebeck coefficient[1], S, and some typical values are

---
[1] see notes on page 216 for comment on choice of terms and symbols

1.2. THERMOELECTRICITY

given in Table 1.2.

Table 1.2: Values of Seebeck coefficient, S (μV K^{-1}), for various metals near 20 and 1000°C. The data from references [12, 13, 14], being coefficients relative to Pt, were corrected using S for Pt, from page 214.

Conductor	Average over 0–100°C	20°C	1000°C	Reference
Chromel		22.2	9.4	[12]
Fe		13.3	−7	[12]
Nicrosil		11.8	8.8	[13]
Au		2.0	4	[15, 16]
Cu		1.9	7	[15, 16]
Ag		1.7		[15, 16]
W		1.3	20.3	[17]
Manganin	0.6			[14]
Brass (Cu50Zn)	0.5			[14]
Phosphor bronze	0.0			[14]
Solder (Sn50Pb)	−0.9			[14]
Stainless steel (18-8)	−1.1			[14]
Pb		−1.0		[15]
Al	−1.3			[14]
Hg	−5.0			[14]
Pt		−4.7	−21.4	[17]
Nisil		−14.8	−29.8	[13]
Alumel		−18.2	−29.6	[12]
Ni		−19.5	−35.4	[18]
Constantan (TN,EN)		−38.3	−65.6	[12]

So far, we have assumed that the temperature gradient was set up instantaneously. If this were possible the exponential decay in cloud drift and the growth of the electric field to its steady state value would have been almost instantaneous. Its response time would be considerably less than the thermal response time for setting up the temperature gradient in a real situation. So, in practice, the electric field keeps up with the temperature gradient as it is established, and also if it continues to change. E and dT/dx are always related according to equation (1.1).

In the above example for copper, a temperature gradient of 100°C cm^{-1} produced a heat flow of 400 W. Under these severe conditions the resultant electric field is rather small. It is 0.00019 V cm^{-1}.

Equation (1.1) is an expression of the Seebeck effect in terms of the electric field, which is the rate at which the electric potential (emf) increases with distance, x, along the conductor, i.e., $E = dV/dx$. Hence

$$\frac{dV}{dx} = S\frac{dT}{dx}$$
$$\text{or} \quad dV = S\,dT. \tag{1.2}$$

In other words, along the elemental length, dx, the temperature difference, dT, gives rise to an emf dV equal to $S\,dT$ (and not dependent on the length, dx!).

Summing along the conductor between any two points, having the temperatures T_1 and T_2, we obtain the net emf developed in the length,

$$V = \int_{T_1}^{T_2} S(x, T)\,dT \tag{1.3}$$

and this represents the Seebeck effect in terms of emf.

In this equation, S is given as a function of both T and x to emphasise that S depends not only on temperature but also varies from point to point (x) along a conductor, as does the chemical composition and other physical properties. In other words, real materials are not homogeneous, and the above integral does not have a unique value: it depends on the temperature profile.

Nevertheless, let us for the moment ignore this complication and consider the ideal case of a homogeneous material—in the next section this will be used as the starting point for delineating thermocouple behaviour.

The Seebeck emf in a length of homogeneous material is produced and distributed along it in response to the temperature profile. Yet, it does not depend on the profile. The emf, from equation (1.3) with S a function of T only, would then be completely independent of the temperature distribution and would have a unique value. Notice also that the equation contains no geometric parameters. Therefore, the Seebeck emf, V, developed within a conductor is not directly dependent on diameter and length.

Put simply, the net Seebeck emf in any given length of homogeneous material, depends on three things only: it depends on the temperature at each end, and it depends on the Seebeck coefficient.

**Emf
is NOT generated
at a junction between two metals.**

1.3 The thermocouple and its emf

Over the years there have been many attempts to describe the principles of operation for a thermocouple. Most have been incorrect, because they have drawn on faulty concepts of tip emf, propagated in the early literature. This deficiency has been highlighted in several articles [6, 8, 9, 19], which offer a more reasonable explanation. What follows is a similar account, differing only in its method of presentation.

To summarise from the previous section:
- the Seebeck effect is a second-order consequence of electrons managing the business of heat flow—an emf is developed wherever heat is flowing.
- In any small region there will be a quantity of emf proportional to the rate at which heat flows through it. As a result, emf will be concentrated in regions of high temperature gradient and be vanishingly small in regions of uniform temperature.

For a **homogeneous** wire, one whose properties don't vary from point to point within it, the distribution of emf along its length will be established as a response to the temperature profile and will change if the profile changes. However, the total emf will not change—provided of course that the temperature of each end remains unchanged. Thus, the signal generated in the wire is a characteristic of the material and is a unique function of the temperatures at its two ends. For an inhomogeneous (real) wire see page 13.

This is the basis for a temperature sensor—a signal produced in a single wire as a consequence of the temperature varying along its length. To measure the signal, however, two other wires would be needed to connect it to an instrument and, of course, they would contribute signals of their own. This form of circuit is treated in section 3.9.

I find it more convenient to consider wires in pairs, experiencing the same temperature profile (Figure 1.3). If they are joined at one end, at a temperature T_1, and at their other ends the temperatures are the same, at T_2, then the net signal, $V_{1\to 2}$, will characterise the pair of materials and still be a function of two temperatures only. To measure the signal a connecting lead comprising two identical wires, of copper for example, is used. Since the two connecting wires operate over the same temperature interval, from T_2 to the temperature of the instrument terminals, they each produce the same Seebeck emf and their net effect is zero. As a result, the signal appearing at the instrument is exactly that produced by the pair of dissimilar wires that represent T_1 and T_2.

The dissimilar pair is a **thermocouple**, each wire is one of its 'thermoelements', and the two 'open' ends at T_2 are referred to collectively as the 'reference junction' or **'cold junction'** (CJ). Its signal, $V_{1\to 2}$, is the difference

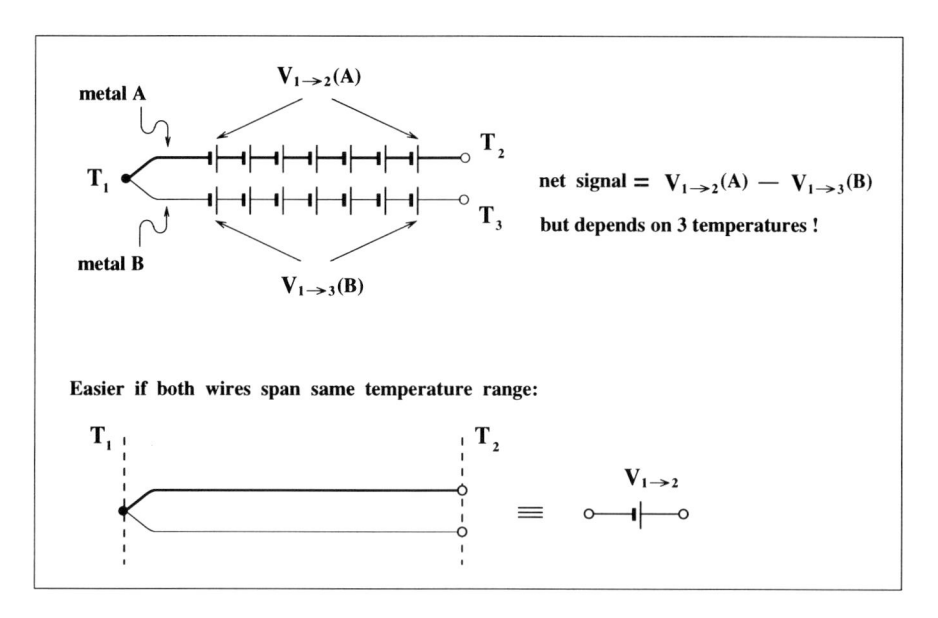

Figure 1.3: The ideal thermocouple: two wires operating between two isotherms, at T_1 and T_2, to produce a characteristic signal, $V_{1\to 2}$, that depends only on these two temperatures.

between the individual signals in its two wires, metal A and metal B, and it is convenient to define a Seebeck coefficient for the thermocouple as the difference $(S_A - S_B)$. Then, equations (1.2) and (1.3), given above for any one material, apply also to the thermocouple, where S is then the net coefficient $(S_A - S_B)$.

On most occasions we can ignore the contributions from individual wires and regard the thermocouple as a single source $V_{1\to 2}$, which reflects the two temperatures, T_1 and T_2, and is driven by a net Seebeck coefficient S, representing the particular pair of metals. This approach is very convenient, but does require that the two connections at the CJ have the same nominal temperature, that $T_2 = T_3$ in Figure 1.3.

If $T_2 \neq T_3$ the temperature interval from T_2 to T_3 would not be represented by emf in both thermocouple elements, metals A and B of Figure 1.3. In one leg of the circuit the interval would occur along one of the thermocouple elements and in the other, along one of the connection wires (usually of copper) that connect the thermocouple to an instrument. So, the net emf for this interval is spurious—it neither corresponds to the behaviour of the thermocouple pair nor is it zero, that relevant to the connecting wires (Cu versus Cu). The difference $T_2 - T_3$ contributes error.

In practice (Chapter 3) a thermocouple circuit may not be as simple in form

1.3. THE THERMOCOUPLE

as just described and will thus be more interesting, because of the greater number of potential error sources (section 5.7). There are also questions relating to the choice of conductor materials and the method of insulating them from each other that need to be answered (section 2.10).

1.3.1 Emf distribution

The thermocouple tip is important: it needs to survive mechanically and it must maintain good electrical contact between the two conductors, but, it plays no part in producing Seebeck emf.

The notion of emf being distributed along thermocouple wires needs to be examined further to assess all its ramifications. To begin with, Seebeck emf always accompanies heat flow within each conductor and is proportional to it. Its distribution is thus dictated by the temperature profile, which can usually be represented as shown in Figure 1.4. Here, the thermocouple, in bridging the temperature step from ambient at its CJ to the temperature at its tip, experiences most of the change in a small part of its length, in the vicinity of a furnace wall. The two remaining sections, within the furnace hot zone and at ambient, are nominally uniform in temperature (or should be) and hence produce but a small part of the total signal. So, most of the thermocouple emf is concentrated in one region, at the furnace wall: the region of greatest temperature gradient or heat flow.

The distribution of emf in a homogeneous thermocouple is of no consequence. What matters is that temperature varies continuously along the length of a thermocouple from tip to CJ, that all this change is represented by appropriate emf and that the total emf in a homogeneous thermocouple is a unique function of its tip and CJ temperatures, from equation (1.3).

For real thermocouples, however, the position is not this simple. The emf is not just a function of the two temperatures—it has a second-order dependence on the temperature profile and the distribution of emf becomes important. In real thermocouples attention must focus on the temperature gradient regions, as will now be discussed.

1.3.2 Inhomogeneity

Real thermocouples are inhomogeneous when new and, with use, the level of inhomogeneity increases. The initial level of inhomogeneity is small but significant. For Pt-based thermocouples it would be, typically, ±0.02% and for base-metal alloys it may be as small, if the wires are kept below 200°C. At higher temperatures, the as-new level for base-metal thermocouples is about ±0.1%. These magnitudes of inhomogeneity reflect the degree to which a

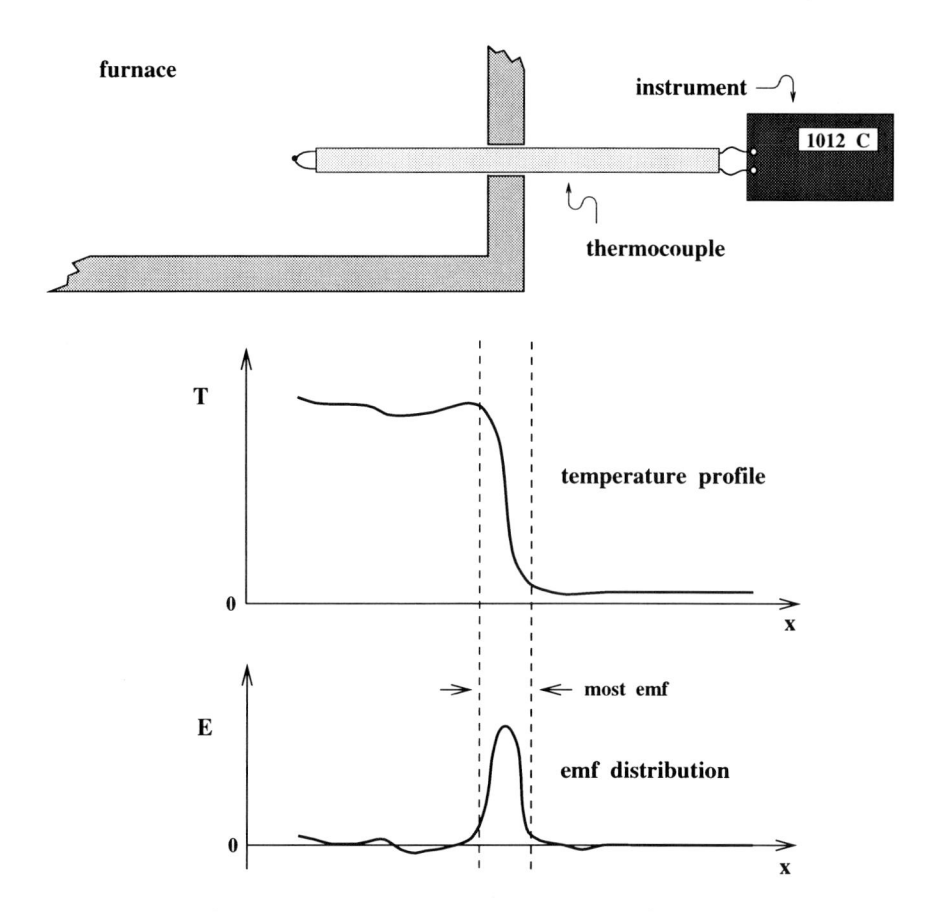

Figure 1.4: Emf distribution in a thermocouple: a plot of its temperature, T, and the resultant electric field, E (emf/unit length), along its length, x.

thermocouple's output depends on its longitudinal temperature profile. In other words, if the tip and CJ temperatures of a thermocouple are somehow held constant while its depth of immersion into a furnace is progressively changed its output would vary to this extent.

The Seebeck coefficient is a measure of how the electrons are coupled to their environment—to the metal lattice and, on the larger scale, to the grain structure. It is sensitive to changes in the chemical and physical features of the metal and, thus, will change if the metal is contaminated, oxidised, strained or heat treated etc. Since all these processes depend on temperature and since a thermocouple, by virtue of its mode of use, necessarily experiences a non-uniform temperature field, a non-uniform change in Seebeck coefficient occurs. The wires thereby acquire a thermoelectric imprint or 'signature' that characterises the particular thermal history being accumulated. By signature

1.3. THE THERMOCOUPLE

I mean the longitudinal profile of the Seebeck coefficient, i.e., a plot showing the coefficient as a function of distance along the thermocouple. A practical means of measuring the signature (inhomogeneity) is to effectively calibrate the thermocouple at one temperature as a function of depth of immersion (see page 126). For new wires the signature would have random fluctuations along it by as much as $\pm 0.1\%$, as suggested above. For the sake of clarity, however, such variation is ignored in Figure 1.5.

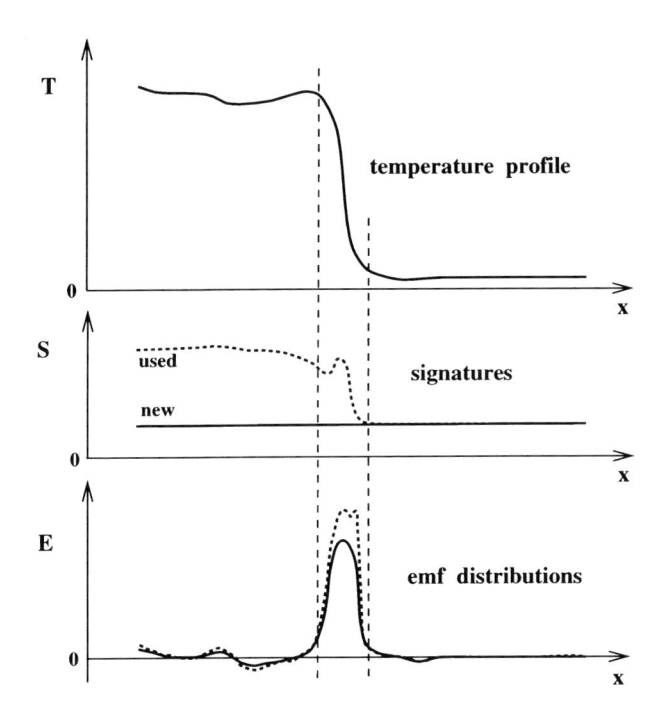

Figure 1.5: The change in thermoelectric signature with use and the resultant change in the emf distribution. Plotted are the temperature, T, the Seebeck coefficient, S, and the electric field, E—for convenience, S for new wire is assumed independent of T.

Figure 1.5 illustrates the growth in signature that normally occurs in thermocouples kept at a fixed immersion and used for temperatures above about 250°C. The effect of this change on the emf distribution is also shown and, clearly, changes in emf are significant only where the temperature gradient is high. The thermocouple experiences a long-term drift in calibration that does not depend significantly on the shape of the temperature profile even though the drift is produced as a response to it. New thermocouples, cut consecutively from the same reel and placed in different furnaces at the same temperature, would have similar initial calibrations and would drift by similar amounts. The emf distributions and the signatures would of course differ, depending on the shape of the particular temperature profiles.

1.3.3 Changes in immersion

Consider now the question of moving a thermocouple to different depths of immersion in the same furnace, with the temperatures of the tip and CJ held constant. For new wires we have already noted that small fluctuations in signal will occur (up to ±0.1%) as the immersion is changed. For a used thermocouple the changes are greater, especially if it has been at the one immersion for enough time to develop a noticeable drift in calibration.

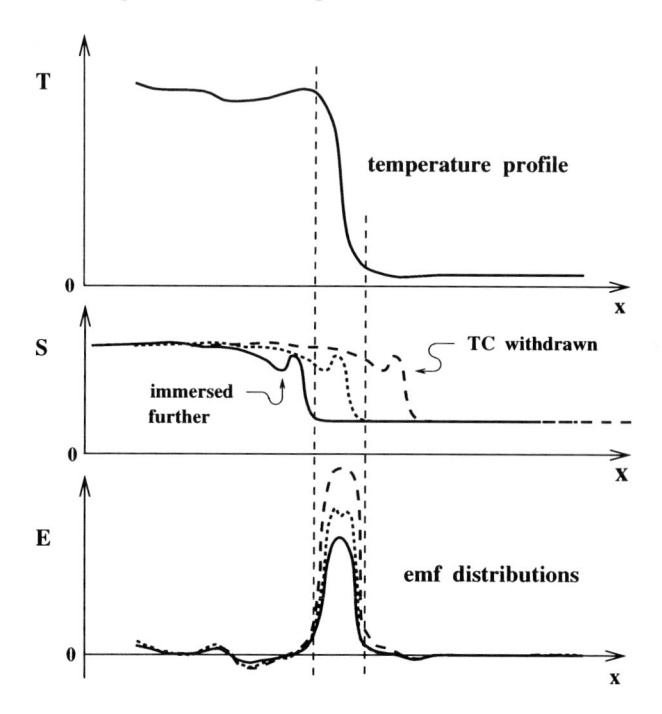

Figure 1.6: The result of moving a thermocouple relative to the temperature profile after the use depicted in Figure 1.5. The emf distribution is shown for three positions of the thermocouple, each represented by the signature.

The effect of moving a used thermocouple is illustrated in Figure 1.6. Here, the signature had been developed in a previous use at the intermediate immersion and on moving the thermocouple different parts of the signature are brought into the critical temperature gradient zone, so affecting the level of emf. If the depth is increased, so as to place only unused wire in the gradient region, the calibration will revert to its 'as-new' value and the signal will rapidly fall by the amount it had drifted up over the previous period of use (taking some minutes, depending on the response time of the probe). Of course, if the thermocouple is now held in this position for a time comparable to that of its previous use the sorts of change evident in the signature of Figure 1.5 will develop, 'adding' to that already produced by the earlier use.

Alternatively, the signal will increase if the immersion is decreased, particularly if the main temperature-gradient region is now within the section of wire most affected by the earlier history. The change in calibration, resulting from reducing the immersion of a used thermocouple, can be considerable: an order of magnitude greater than the *in-situ* drift that occurred before the move. For example, 0.5 mm (24 AWG) nickel-based thermocouples would drift only about 1°C in 16 h, while monitoring a temperature of 1000°C, yet, if their immersions are decreased their calibrations could change a further 15°C [20].

Clearly, thermocouples work reasonably well if fixed in position relative to the temperature profile. They will then drift in calibration, but the change will be gradual and is easily monitored and corrected. To move a thermocouple in ignorance could be disastrous even if, on relocating it in another furnace, the depth of immersion is nominally the same. Examples of this are given on pages 21 and 21.

1.3.4 Laws of thermoelectric circuits

It is usual, in discussing thermocouples and their circuits, to include an analysis using the three 'laws of thermoelectric circuits' [10, 11]. The 'laws' are simple empirical rules of thumb, useful in the days when users were surrounded by conflicting explanations for thermocouple emf, each being based, at least in part, on the faulty notion of tip emf. The laws are self evident, once the source of thermocouple emf is understood. Indeed, their only use is as an alternative to such understanding, and for this reason they are not detailed here.

1.4 Standard reference tables for thermocouples

Variations in the degree and types of impurity in a metal and variations in the chemical composition of an alloy affect changes in the Seebeck coefficient, and thus its emf. Some controls are thereby needed during manufacture before the thermocouple can be considered a practical temperature sensor. This is done by requiring thermocouple manufacturers to have each alloy, or pair of alloys, conform to a defined relationship between emf and temperature, within specified tolerances (see page 27). Such relationships are expressed as polynomials, the 'standard reference functions' (Appendix B), and often given in tabular form. The standard reference tables, such as those given in references [21, 22], contain values of emf at 1 or 10°C intervals for each of the popular thermocouple types. For convenience, an abbreviated set of reference tables is given in Appendix B, beginning on page 228.

Table 1.3: Fragments from the standard reference tables for type N thermocouples. Data are values of thermocouple emf, in μV, for various tip temperatures, assuming a cold junction of 0°C (see also page 232).

°C	0	1	2	3	4	5	6	7	8	9
0	0	26	52	78	104	130	156	182	208	235
10	261	287	313	340	366	392	419	445	472	498
20	525	552	578	605	632	659	685	712	739	766
30	793	820	847	874	901	928	955	983	1010	1037
40	1065	1092	1119	1147	1174	1202	1229	1257	1284	1312
..	..	..	..	..	..	..	..	..	..	..
..	..	..	..	..	..	..	..	..	..	..
500	16748	16786	16824	16863	16901	16939	16978	17016	17054	17093
510	17131	17169	17208	17246	17285	17323	17361	17400	17438	17477
..	..	..	..	..	..	..	..	..	..	..
..	..	..	..	..	..	..	..	..	..	..
800	28455	28494	28533	28572	28612	28651	28690	28729	28769	28808
810	28847	28886	28926	28965	29004	29043	29083	29122	29161	29200
..	..	..	..	..	..	..	..	..	..	..
..	..	..	..	..	..	..	..	..	..	..
1100	40087	40125	40163	40201	40238	40276	40314	40352	40390	40428
1110	40466	40504	40542	40580	40618	40655	40693	40731	40769	40807
1120	40845	40883	40920	40958	40996	41034	41072	41109	41147	41185
1130	41223	41260	41298	41336	41374	41411	41449	41487	41525	41562
1140	41600	41638	41675	41713	41751	41788	41826	41864	41901	41939
1150	41976	42014	42052	42089	42127	42164	42202	42239	42277	42314

The thermocouple user is interested in converting Seebeck emf into a value of temperature, and there are two options. Firstly, it can be done by an instrument or computer, using a suitable functional relationship (e.g. the standard reference function), or, secondly, the conversion may be achieved by manual application of the standard reference tables. For further comment see page 221.

Consider Table 1.3, which shows a part of the standard reference tables for type N thermocouples[2]. Note in particular that the data apply only to thermocouples with a CJ temperature of 0°C. If the CJ is other than at 0°C a further calculation is required (see page 19). But first, let's examine the table. In the row beginning 800°C and the column headed 2°C is the value 28 533 μV, which is the reference emf for a type N thermocouple corresponding to 802°C (tip at 802°C and CJ at 0°C). If the tip temperature of the thermocouple is increased 1°C to 803°C its signal should increase to 28 572 μV, given in the

---
[2] See section 2.2 for a description of this and other thermocouple types.

1.4. STANDARD REFERENCE TABLES

next column. The increase of 39 μV for a 1°C change is the Seebeck coefficient near 803°C since, from equation (1.2), $S = dV/dT$. Notice, using the top row of the table, that at 0°C the Seebeck coefficient is somewhat less, being 26 μV K^{-1}.

In practice, it would be unlikely for the thermocouple emf to equal one of the values in the table. Simply selecting the nearest tabulated value would give T to the nearest ± 0.5°C and for better resolution interpolation is required. This is done by assuming that temperature changes linearly with emf, between adjacent values in the table: a safe assumption for tabulations given in increments of 10°C or less (see page 221). For example, a measured emf of 17 220 is 12 μV above 17 208, the value for 512°C. At this temperature a 1°C increase in tip temperature would cause a 39 μV change. Thus, 12 μV is equivalent to 12/39 or 0.3°C and the temperature equivalent to 17 220 μV is 512.3°C.

1.4.1 The use of reference tables when CJ $\neq$ 0°C

The thermocouple of Figure 1.7, having a tip temperature of T and a CJ at 0°C, develops a signal $V_{T \to 0}$ that can be fed directly into the reference tables to obtain a value of T. This is also true of any part of the thermocouple adjacent to and including those points at 0°C, in this case the CJ, for example that section from 30 to 0°C. It produces the emf $V_{30 \to 0}$, which is 793 μV for the type N thermocouple.

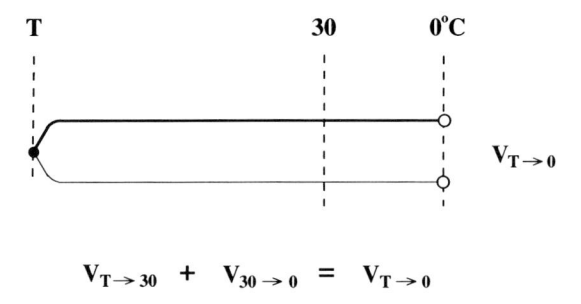

Figure 1.7: Breakdown of thermocouple signal to illustrate method of using reference tables for a CJ temperature other than 0°C.

If the open ends of the thermocouple, its CJ, were heated from 0 to 30°C, or if measurement leads were attached to the wires at the 30°C isotherm, the measured emf would be reduced to $V_{T \to 30}$, smaller by $V_{30 \to 0}$. To use the reference tables $V_{T \to 0}$ is required. Clearly, this is $V_{T \to 30} + V_{30 \to 0}$, or in general,

$$V_{T \to 0} = V_{T \to CJ} + V_{CJ \to 0}. \tag{1.4}$$

For example, consider a signal of $28\,490\,\mu\text{V}$ developed in a thermocouple with a CJ at 27°C. The value 28 490 should not be fed directly into the reference tables, because it is inappropriate. To do so would yield 800.9°C, and then if the CJ temperature were added, as some users tend to do, a temperature of 827.9°C is obtained. Compare this with the correct result, available from equation (1.4) as follows. The measured value, 28 490, represents $V_{T\to 27}$ and we need $V_{T\to 0}$. To get this, we add to $V_{T\to 27}$ the 'missing' signal, $V_{27\to 0}$, which is $712\,\mu\text{V}$ (Table 1.3). Hence $V_{T\to 0}$ is $29\,202\,\mu\text{V}$ and $T = 819$°C (Table 1.3).

1.4.2 Use of reference tables in diagnosis

Sometimes, there is a concern about the contribution from a particular part of the thermocouple circuit. It could, for example, be contaminated or contain a switch or some other foreign object. When this happens some useful estimates can be made. Consider Figure 1.8. Here, $V_{1\to 2}$ is the contribution to the total signal developed in that section of the thermocouple between the isotherms at T_1 and T_2. The emf is dictated by these temperatures and is independent of the circuit outside the section. Its value is the same whether the section remains as part of the thermocouple, as shown on the left of the figure, or removed to form a short thermocouple, experiencing the same temperature profile, as on the right. Equally, $V_{1\to 2}$ is that signal measured when the wires are shorted at T_1 and measuring leads are attached at T_2.

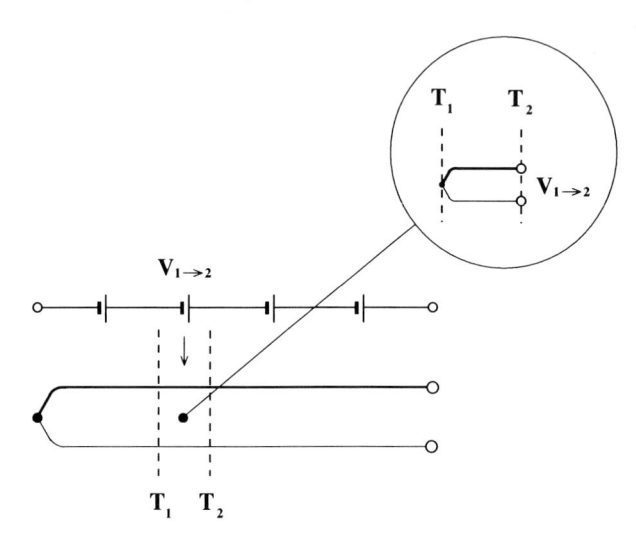

Figure 1.8: Section of thermocouple between isotherms at T_1 and T_2 represented as a separate thermocouple, for the purposes of calculating its contribution to the total signal.

1.4. STANDARD REFERENCE TABLES

Hence, if a section of thermocouple appears to be misbehaving, or is in doubt for some other reason, it can be 'isolated' in this way. Values for $V_{1\to 2}$ may then be obtained from the reference tables, on making suitable assumptions for the Seebeck coefficient (see below) and the temperatures T_1 and T_2.

Suppose a short segment of thermocouple is contaminated by carbon. To look at possible consequences let's assume that the carbon had decreased the Seebeck coefficient by 10% and consider the segment in different, hypothetical situations. If it is placed in a high temperature-gradient zone, to span an interval of maybe 300°C, say from 500 to 800°C, its emf contribution would be 10% lower than it would have been if uncontaminated. A type N thermocouple would thus be low by $0.1 \times (28\,455 - 16\,748)$, or $1\,171\,\mu$V, using values for 800 and 500°C from Table 1.3. Alternatively, the contaminated segment could have been located in the ambient region and span only 10°C, from 25 to 35°C. Its contribution would then be 10% less than $(928 - 659)$, i.e., in error by $-27\,\mu$V.

Thus, in the assumed locations, the contaminated segment causes an error of either $-1\,171$ or $-27\,\mu$V. This is a loss for the total signal, and so its effect on the measured value of temperature, obtained with the thermocouple, is dictated by the Seebeck coefficient at the tip temperature. For a tip near 1130°C we have $S = 38\,\mu$V K^{-1} and the corresponding error [$\delta V/S$, from equation (1.2)] is 31 or 0.7°C, respectively.

Another way of calculating the equivalent temperature error is to begin with an assumed value for the tip temperature, say 1130°C, and then, from Table 1.3, obtain the equivalent emf, $41\,223\,\mu$V. The above error, either $-1\,171$ or $-27\,\mu$V for the two hypothetical situations, would reduce the thermocouple signal to 40 052 or 41 196 μV, respectively. This corresponds to 1099.1 or 1129.3°C (Table 1.3), in other words, a temperature low by 31 or 0.7°C, as found above.

Consider a second example, represented by Figure 1.9. Here, a type N thermocouple had been operated at 1100°C for a few months, without a change in immersion, and the owner suspects that its calibration had drifted significantly during the period. Let us calculate what the drift may have been and what would happen if the thermocouple is calibrated. The change in signature, shown in Figure 1.9, was estimated, using data given in section 2.6, on making certain assumptions about diameter and so on—it is thus a realistic example. Nevertheless, I am not suggesting that the long-term drift of a thermocouple should ever be calculated as a means of correcting for such drift. It is done here, merely to highlight the consequences of removing the thermocouple for calibration and as a further example of analysis using reference tables.

Returning to the figure, we see that S had changed considerably in the hot zone of the furnace and not at all outside it. Neither of these regions contribute significant emf, so we will ignore them to simplify calculation. The section of thermocouple in the temperature gradient region, between about 500 and 1100°C, had been affected—here, the average increase in S was about 1.5%, according to the figure. This section should have produced 40 087 − 16 748, or 23 339 μV, from the reference tables (Table 1.3); and a 1.5% increase is 350 μV, which corresponds to 9.2°C at 1100°C ($S = 38\,\mu\text{V K}^{-1}$). The thermocouple had thus drifted up by about 9°C.

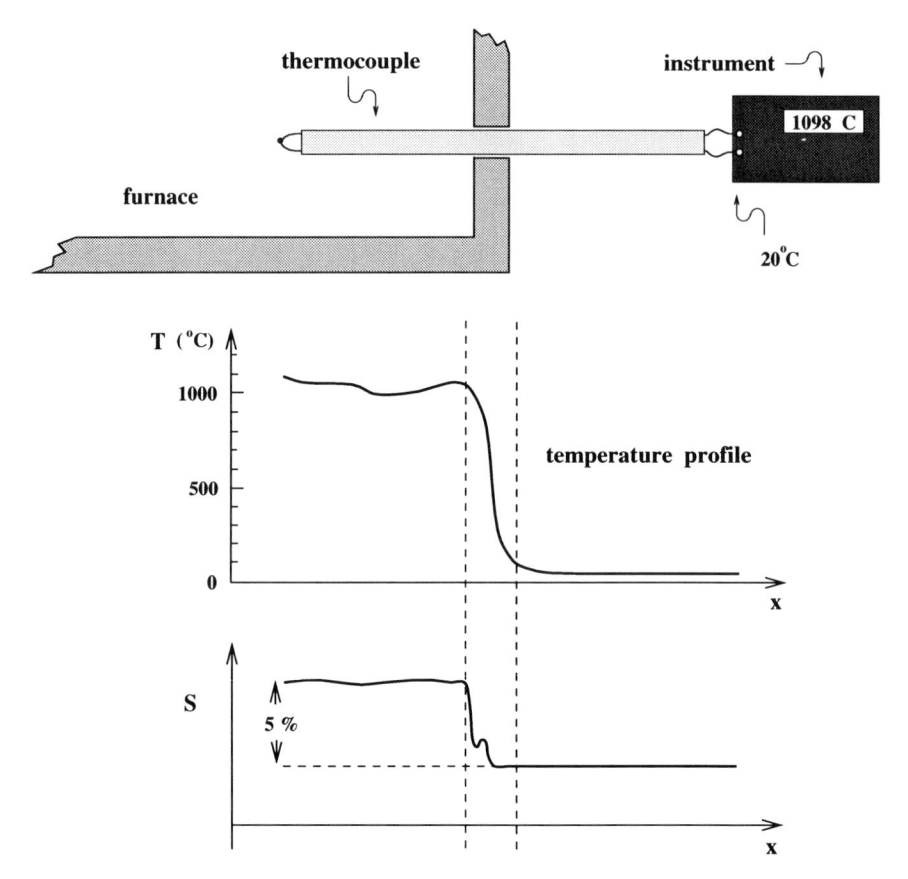

Figure 1.9: Long-term change in a thermocouple used at 1100°C. The suggested change in Seebeck coefficient, S, is used in calculating drift and in predicting the result of calibration—the 5% change is relative to the as-new value of S, assumed for convenience to be independent of T.

But would a calibration reveal this? The thermocouple, placed in a laboratory furnace, would then be in a different temperature profile. To see

1.4. STANDARD REFERENCE TABLES

how significant this change is let's examine two alternative possibilities that are easy to calculate. Consider the temperature gradient zone of the calibration furnace falling completely within either the region least affected by its earlier use or the region most affected. Other positions would yield intermediate results.

In the first case, the calibration would relate to 'unused' wire and produce similar values to those found earlier at the initial calibration, to suggest the thermocouple had not changed. In the second case, the calibration furnace would subject the affected part of the thermocouple to temperatures from ambient, say 20°C, to 1100°C. This corresponds to $40\,087 - 525$, or $39\,562\,\mu V$, from Table 1.3, and so the thermocouple would produce 5%, or $1\,978\,\mu V$, more. Taking the Seebeck coefficient as $38\,\mu V\,K^{-1}$, this represents an error of 52°C.

In other words, a calibration of the thermocouple at 1100°C would suggest a signal that differs from its 'as-new' value by the equivalent of 0 to 52°C, depending on the immersion used. Compare this with 9°C, the actual drift. From this example, two conclusions follow.

- The *in-situ* drift in signal of a thermocouple used for long periods at high temperatures is small compared to the changes taking place in its thermoelectric signature.
- To move the thermocouple for another measurement, or for calibration, could be disastrous. If a thermocouple has had sufficient use for its Seebeck coefficient to have changed and it needs calibration it must be done *in situ* (see section 4.10).

For further comment see item (f) on page 68.

Chapter 2

Thermocouple Materials and their Properties

2.1 Introduction

In this chapter I describe the physical properties of the components of a 'thermocouple assembly', and I make a distinction between two mechanically and behaviourally different thermocouple formats. One is the mineral-insulated, metal-sheathed (**MIMS**) system, in which the thermoelements, their insulation and a sheath are integrated into a flexible cable. The other format is loosely referred to as the **bare-wire** thermocouple because above about 500°C the thermoelements are usually insulated in loose-fitting beads or a woven high-temperature material. The wires are thus exposed to the local atmosphere and are vulnerable to the effects of oxygen, carbon and sulphur, etc. At lower temperatures other forms of insulation are available, which effectively seal in the wires, and they are no longer 'bare'. Nevertheless, I still refer to them as bare-wire thermocouples—a convenient means of distinguishing them from the MIMS varieties (section 2.6.3). The insulation and sheathing of bare-wire thermocouples are discussed separately in section 2.10.

2.2 Conventional thermocouple types

There are something like 300 different types of thermocouple for which information is available [5]. Over 20 combinations are extensively used and, of these, 8 have been standardised, with their emf-temperature relationships represented by internationally recognised standard reference functions (Appendix B). The metals used in the standardised thermocouple types are given in Table 2.1 with the letters, B, S, ..., E, that identify them. The net Seebeck coefficients for the common thermocouples are given in Table 2.2.

The practice of using letters to designate thermocouple types was originated by the Instrument Society of America and adopted in 1964 as American National Standard C96.1 (later revised as ANSI-MC96.1). Apart from the eight types covered by ANSI, there are three others sufficiently used to warrant mention. They are the types L and U, to describe thermocouples made to pre-1990 DIN standards for alloys similar to, but significantly different from, types J and T, respectively, and type C for W 5Re versus W 26Re (page 72).

Table 2.1: Thermocouple types in common use: their letter designations and typical compositions (wt.%).

Type	Thermoelements Positive	Negative	Composition Positive	Negative
B	Pt 30Rh	Pt 6Rh	Pt 29.6±0.2Rh	Pt 6.12±0.02Rh
R	Pt 13Rh	Pt	Pt 13.00±0.05Rh	~99.99%Pt
S	Pt 10Rh	Pt	Pt 10.00±0.05Rh	~99.99%Pt
K	Chromel[†]	Alumel[†]	Ni 9.5Cr 0.5Si	Ni 5(Si,Mn,Al)
N	Nicrosil	Nisil	Ni 14.2Cr 1.4Si	Ni 4.4Si 0.1Mg
E	Chromel	Constantan	Ni 9.5Cr 0.5Si	Cu 44Ni
J	Iron	Constantan	~99.5%Fe	Cu 44Ni[‡]
T	Copper	Constantan	~99.95%Cu	Cu 44Ni

[†] registered trademarks of the Hoskins Manufacturing Co.
[‡] Not interchangeable with the Constantan of types E or T.

Basically, the thermocouples of Table 2.1 fall into three groups. The first is the rare-metal group, comprising types B, R and S, based on platinum and its alloys with rhodium and discussed further in section 2.5. They are the most accurate of the tabulated thermocouples and may be used at higher temperatures, but, they are more expensive and are particularly sensitive to contamination.

The second group consists of the two nickel-based thermocouples, types N and K. They are preferred for most applications not requiring the higher temperature limit or accuracy of the rare-metal thermocouple. Of the two types, N and K, the best choice depends on several things, and especially on whether they are in the bare-wire or the MIMS form. When the thermoelements are bare the type N has distinct advantages (section 2.6.2) and for MIMS probes, the position is less clear (see page 67).

The third group, comprising the types E, J and T (section 2.7), is based on

2.2. THERMOCOUPLE TYPES

the use of Constantan as the negative leg. Constantan has the most negative value of Seebeck coefficient (Figure 2.2), so the net Seebeck coefficients for the third group are high.

The thermocouples of Table 2.1 cover the temperature range from about 30 K to ~1750°C. Other thermocouple types, described in section 2.8 (page 70), extend the usefulness of thermocouples beyond these limits.

Table 2.2: Some values of net Seebeck coefficient, S (μV K^{-1}), for the common thermocouple types. Values of thermocouple emf for temperatures incrementing by 10°C are given in Appendix B, beginning on page 228.

Temperature (°C)	\multicolumn{7}{c}{S for Thermocouple Type}
	B	R	S	K	N	E	J	T
−200				15.3	9.9	25.1	21.9	15.7
−100				30.5	20.9	45.2	41.1	28.4
−50		3.7	4.0	35.8	24.3	52.6	46.6	33.9
0	−0.2	5.3	5.4	39.5	25.9	58.7	50.4	38.7
50	0.3	6.5	6.5	41.2	27.7	63.2	52.8	42.8
100	0.9	7.5	7.3	41.4	29.6	67.5	54.4	46.8
200	2.0	8.8	8.5	40.0	33.0	74.0	55.5	53.1
300	3.0	9.7	9.1	41.4	35.4	77.9	55.4	58.1
400	4.1	10.4	9.6	42.2	37.1	80.1	55.2	61.8
500	5.0	10.9	9.9	42.6	38.3	80.9	56.0	
600	6.0	11.4	10.2	42.5	39.0	80.7	58.5	
700	6.8	11.8	10.5	41.9	39.3	79.7	62.2	
800	7.6	12.3	10.9	41.0	39.3	78.4		
1000	9.1	13.2	11.5	39.0	38.6	75.2		
1200	10.4	13.9	12.0	36.5	37.2			
1400	11.3	14.1	12.1					
1600	11.7	13.9	11.9					

2.2.1 Manufacturing tolerances

Various Standards organisations throughout the world have specified tolerances for the manufacture and supply of thermocouple materials. These relate to the initial (as-received) output of thermocouples and endeavour to control the extent of deviations from the standard reference tables (section 1.4)—usually given as the equivalent differences in temperature. Currently, the magnitudes of the recommended tolerances in most countries are much the same. This was not always the case and alloy producers have had to supply to a variety of tolerance levels. For use above 0°C thermocouples are produced in one of two tolerance categories, referred to as 'standard tolerance' and 'special tolerance', or 'class 1' and 'class 2', respectively. Such thermocouples are often described as having standard-grade and premium-grade wires. Thermocouples are also produced explicitly for sub-zero temperatures, and for these, different tolerances apply. Furthermore, those thermocouple alloys produced as 'extension wires' (section 3.4) have even broader tolerances.

Table 2.3: Manufacturing tolerances for conventional thermocouple types [22]. Note that tolerances apply only to a limited temperature range.

Type	Temperature Range (°C)	Standard Tolerance* (°C)	Special Tolerance* (°C)
B	870 to 1700	±4.4 or ±0.5 %[†]	±2.2 or ±0.25%
R or S	0 to 1450	±1.5 or ±0.25%	±0.6 or ±0.1 %
K or N	0 to 1250	±2.2 or ±0.75%	±1.1 or ±0.4 %
E	0 to 900	±1.7 or ±0.5 %	±1.0 or ±0.4 %
J	0 to 760	±2.2 or ±0.75%	±1.1 or ±0.4 %
T	0 to 350	±1.0 or ±0.75%	±0.5 or ±0.4 %
E	−200 to 0	±1.7 or ±1.0%	
K	−200 to 0	±2.2 or ±2.0%	
T	−200 to 0	±1.0 or ±1.5%	

[†] the percentage applies to the tip temperature in °C and assumes a CJ at 0°C.
* whichever of the two values below is the larger.

Recommended tolerances for conventional thermocouple types, as expressed in reference [22], are given in Table 2.3. Their intent is to minimise the as-new variability in emf from lot to lot for each thermocouple type—they do not indicate performance:

- The different tolerances given for any one thermocouple type are more a reflection of the ease with which alloy producers can adjust or control their emf-temperature relationships than of their performance as temperature sensors.

- A standard grade thermocouple will be just as stable as one produced to a special tolerance. Indeed, a thermocouple that fails to comply with the widest tolerance will drift as little, or as much, as one that complies with the most stringent.

- The specified tolerances refer to new, as-received wire. As seen in the following pages, all thermocouples drift in calibration when used at an elevated temperature—they are likely to drift outside the tolerance limits, given sufficient time and/or temperature.

2.2. THERMOCOUPLE TYPES

Table 2.4: The melting point, electrical resistivity (ρ_{20}) and thermal conductivity (κ_{20}) at 20°C and the linear expansion on heating from 20 to 1000°C for various materials, from references [23, 24, 25, 26, 27, 28, 29].

Material	MP (°C)	ρ_{20} ($\mu\Omega$ cm)	κ_{20} (W m^{-1}K^{-1})	Expansion (%)
Hg	-39	98	8	
Al	660	2.7	237	2.6[†]
Au	1064	2.2	318	1.70
Cu	1083	1.7	401	2.00
Constantan	1220	49	20	
Nisil	1340	37	24	1.46
Alumel	1400	29	27	1.55
Nicrosil**	1410	96	13	1.53
Chromel	1427	71	18	1.64
Ni	1455	6.9	91	1.60
Fe	1537	9.6	80	1.2
Pd	1553	10.5	72	1.43
Pt	1768	10.5	72	1.00
Pt 6Rh	1826	17.5		0.95
Pt 10Rh	1850	18.9	36	0.90
Pt 13Rh	1860	19.6	35	0.90
Pt 30Rh	1927	19.0		0.95
Rh	1963	4.5	150	1.07
Ir	2446	5.3	147	0.75
W 26Re	3120	28.3		0.55
W 5Re	3350	11.6		
W	3414	5.3	173	0.42
304 stainless	$\sim$1420	80	15	2.0
Inconel[‡] 600	$\sim$1400	100	15	1.7
Dense MgO*	$\sim$2900	$\sim 10^{20}$	40	1.4
SiC*	$\sim$2700		30	0.5
Dense Al$_2$O$_3$*	$\sim$2040	$\sim 10^{20}$	30	0.82
Mullite*	$\sim$1830		4	0.6
Fused SiO$_2$*	$\sim$1720	$\sim 10^{20}$	1.3	0.04
Pyrex glass	[††]820	$\sim 10^{20}$	1.0	0.32[†]
Glass wool			0.04	
N$_2$ @ 1 atmos.			0.026	

[†] extrapolated beyond MP to allow comparison.   [††] softening point.
* ceramics for sheathing are discussed on page 78.   [‡] see page 77.
** tabulated data also apply to Nicrosil-plus and Nicrobell-A.

2.3 Properties of conventional thermoelements

Some physical properties of the metals used as conventional thermoelements are tabulated on page 29, and data on other materials have been included for comparison.

Notice that the thermal conductivity, κ_{20}, varies over a range of 30:1 for the metals and that the elements have higher values than the alloys. Hence, for the standardised thermocouple combinations, the sum of the κ_{20} values for the thermoelements is highest (420 W m^{-1}K^{-1}) for type T, containing copper, and an order of magnitude lower for types E, K and N, each of which use two alloy legs. In the cryogenic region, where thermal conduction could be a serious problem, the ratio of the conductivities of copper and the alloys is higher (section 2.8).

The tabulated thermal expansion data are useful when selecting the sheath and thermoelement combination of a MIMS system (see section 2.6.5). Of particular interest are the nickel-based thermoelements and the common sheathing alloys, 304 stainless steel and Inconel 600, yet there is a paucity of reliable expansion data for these materials. Data for various 304 stainless steel specimens [24] scatter by ±10% about the value given in Table 2.4 and the value given for Inconel 600 is the average of those in references [24, 25]. The tabulated values for Nicrosil and Nisil were calculated from coefficients of linear expansion given for the range 400 to 1100 K [27] and extrapolated. The values are a little lower (~0.1%) than those shown for Chromel and Alumel, respectively. Being similar alloys, their expansion coefficients ought to be much the same, as is suggested by data for the smaller temperature range 20 to 100°C [28, 29].

Figures 2.1, 2.2 and 2.3, and Tables 2.5 and 2.6 give the Seebeck coefficients as a function of temperature for the various thermoelements. Included are data for Ni, since it is the major component in four of the thermoelement alloys. The data for Ni and Cu were taken from references [18] and [15], respectively, and those for Pt from section A.5 (page 211). For the others their absolute Seebeck coefficients were calculated by adding the coefficient for Pt to their coefficients given relative to Pt [12, 13].

It is clear from the figures that Chromel has the most positive Seebeck coefficient and Constantan the most negative, which accounts for the type E thermocouple having the largest net Seebeck coefficient. Also apparent, is a discontinuity in the coefficient for Fe near 910°C, where its crystal lattice undergoes an $\alpha - \gamma$ transformation. The transformation affects other physical properties, such as the electrical and thermal conductivities and the coefficient of thermal expansion. Indeed, Fe contracts about 0.2% on heating through the transformation.

2.3. GENERAL PROPERTIES

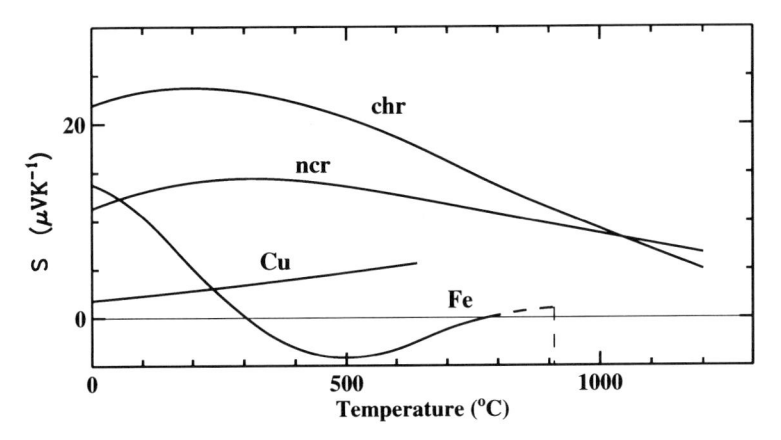

Figure 2.1: Seebeck coefficients, S, of base metals having a positive coefficient (mainly): copper (Cu), iron (Fe), Chromel (chr) and Nicrosil (ncr). Data for Fe has a discontinuity (see text).

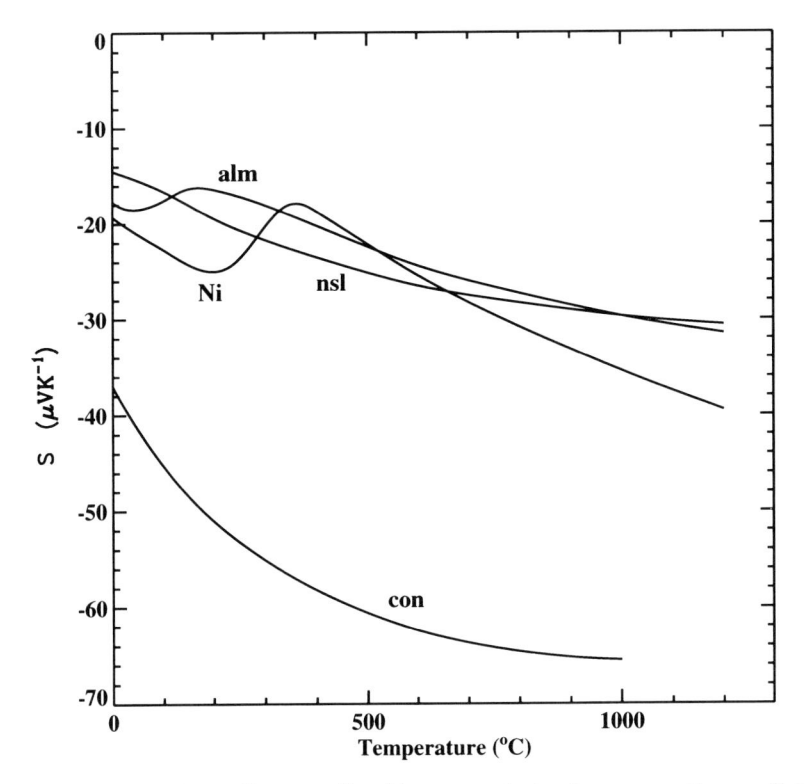

Figure 2.2: Seebeck coefficients, S, of base metals having a negative coefficient: nickel (Ni), Alumel (alm), Nisil (nsl) and Constantan (con).

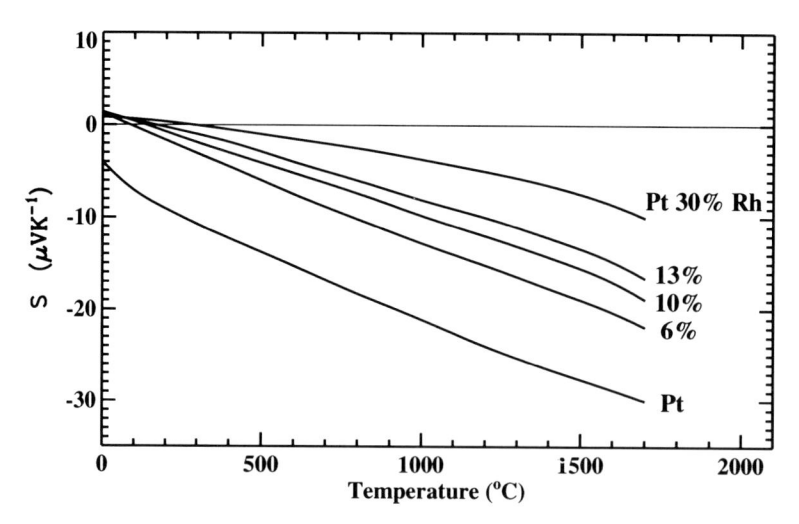

Figure 2.3: Seebeck coefficients, S, of platinum and its alloys with rhodium, as a function of temperature (data are tabulated on page 33).

The Seebeck coefficient peaks at around 370°C for nickel and 180°C for Alumel, a high-Ni alloy, just above their respective Curie temperatures (Figure 2.2). Nisil is a high-nickel alloy, yet its corresponding peak is not immediately obvious in the figure because its higher Si content has pushed its Curie temperature, T_c, below that for Alumel. The Seebeck coefficient is stable near T_c. The peak merely affects the fitting of thermocouple reference equations (Appendix B) and, during calibration, may require smaller temperature steps to minimise interpolation errors.

Table 2.5: Seebeck coefficients (μV K^{-1}) of nickel-based thermocouple alloys. The data are compared with the coefficients of other metals in figures 2.1 and 2.2.

Temperature (°C)	Chromel	Nicrosil	Alumel	Nisil
0	21.8	11.4	−17.7	−14.5
200	23.7	14.0	−16.2	−19.0
400	22.2	14.0	−20.0	−23.1
600	18.5	12.8	−24.0	−26.2
800	13.8	10.9	−27.2	−28.4
1000	9.4	8.8	−29.6	−29.8
1200	5.0	6.6	−31.5	−30.6

2.4. ELEMENTAL THERMOCOUPLES

Table 2.6: Seebeck coefficients (μV K^{-1}) of platinum-rhodium thermocouple alloys.

Temperature (°C)	Pt	Pt 6Rh	Pt 10Rh	Pt 13Rh	Pt 30Rh
0	−4.0	1.0	1.4	1.3	0.8
200	−9.0	−1.8	−0.6	−0.2	0.2
400	−12.3	−4.7	−2.8	−2.0	−0.6
600	−15.2	−7.4	−5.1	−3.9	−1.4
800	−18.3	−10.2	−7.5	−6.0	−2.6
1000	−21.4	−12.9	−9.9	−8.2	−3.8
1200	−24.1	−15.4	−12.1	−10.2	−5.1
1400	−26.5	−17.8	−14.4	−12.4	−6.6
1600	−28.8	−20.5	−17.0	−14.9	−8.8

2.4 Elemental thermocouples

Elemental thermocouples, those whose thermoelements are single elements, are not new and various combinations were reviewed in 1973 [5]. Interest in their potential was revitalised when it was suggested [30] that the accuracy achievable with a Au versus Pt thermocouple was comparable to that of high-temperature Pt resistance thermometers (HTPRT) above about 500°C. A detailed review of the recent position for elemental thermocouples is given in reference [31] and summarised below.

A pure element is thermoelectrically more homogeneous and thermoelectrically more stable than an alloy, because it is free of effects that arise from there being more than one species of atom present, such as lattice ordering and selective volatilisation and oxidation. The most suitable of the elements for use in precision high-temperature thermocouples are the noble metals. They consist of gold (Au), silver (Ag) and the metals of the platinum group—ruthenium (Ru), osmium (Os), rhodium (Rh), iridium (Ir), palladium (Pd) and platinum (Pt). The melting point of Ag (961°C) is too low for consideration in this context and Ru and Os are too hard and brittle to be of use.

The Seebeck coefficients of the metals Au, Pd, Rh and Ir are given relative to that of Pt in Figure 2.4—combinations that have usefully large relative coefficients—about 6 μVK^{-1} at 0°C to $\sim$ 20 μVK^{-1} at 1000°C.

The most stable thermocouple is the Au/Pt, although its use is limited to temperatures below about 1000°C. Under optimum conditions the *in-situ* drift of such a thermocouple seems to be less than 0.01°C in 1000 h at 963°C [32]. The effect of an isothermal anneal for 100 h at 1000°C is also less than

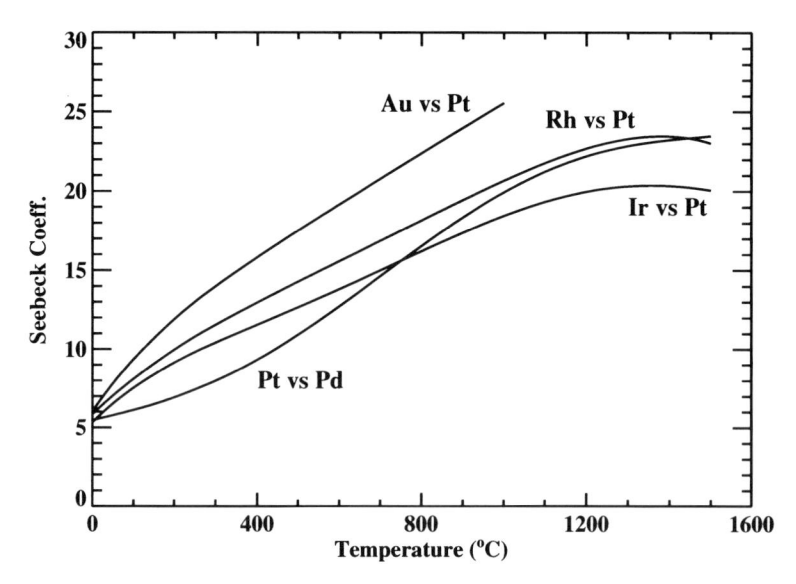

Figure 2.4: The relative Seebeck coefficient (μV K^{-1}) of various elemental thermocouples, based on Pt, from ref. [31].

0.01 °C [30]. By contrast, the instabilities in Pt/Pd are greater, due mainly to the Pd wire. For example, *in-situ* drift in a Pt/Pd thermocouple may be as low as 0.02 °C for 200 h at 963 °C [33], although isothermal annealing causes reversible changes in the Seebeck coefficient of Pd, peaking at around 700 °C by the equivalent of ∼0.1 °C in 200 h [33]. Even so, these changes in Pt/Pd are about 10 times smaller than would occur in a type R or S thermocouple under the same conditions (see section beginning on page 36). No stability data are available for the other elemental thermocouples mentioned above, nor for Pt/Pd above 1100 °C.

To achieve these levels of stability consideration must be given to a variety of issues, e.g.,

- purity,
- annealing procedures,
- relative thermal expansion between the thermoelements and between either thermoelement and the alumina twin-bore insulator (see tabulated data on page 29)—for example, a fine-wire expansion coil is required at the tip of a Au/Pt thermocouple to avoid the related stress,
- the need to constrain wires to avoid stresses in handling and
- oxidation, especially of Ir, Pd and Rh.

These issues are currently under investigation.

2.5 Platinum-based thermocouples (types B, R and S)

Pt-based thermocouples are used for precision laboratory measurements and for industrial applications above 1000°C, where long life and minimal drift are desired. They are potentially the most stable of the conventional types, yet they are the most sensitive to contamination, especially the Pt leg of the type S and type R thermocouples. For example, the effect on either type of 0.1 wt.% additions of Fe, Cr, Si, Ni, Rh or Cu to the pure Pt leg is to change its average Seebeck coefficient for the range 400 to 1100°C by 1.7, 0.75, 0.64, 0.33, 0.21 or 0.15 μV K^{-1}, respectively. To put these changes into perspective, notice that a change of 0.1 μV K^{-1} represents a loss of 1% in the output of a type S thermocouple. Virtually all significant changes occurring in the Pt-based thermocouples can be traced to contamination.

These thermocouples should not be used in reducing atmospheres, nor in the presence of metallic vapours (such as Pb and Zn), non metallic vapours (such as As, P and S) or easily reduced oxides. The effects of low oxygen partial pressure (gettered argon, vacuum etc) on Pt in contact with various refractory oxides, such as Al_2O_3, are described in reference [34]. Platinum-based thermocouples are most reliable in a clean oxidising atmosphere, such as air. It is necessary to remove all traces of lubricating oils, drawing compounds and other sulphur-bearing compounds from the wires and from other metallic components in the thermocouple assembly. Carbon and sulphur are evolved during the decomposition of oils and grease. To clean the wires and metal sheath (inner surface), a nitric acid wash followed by an alcohol rinse may be used; or a wipe with acetone then alcohol (see page 122).

Values of Seebeck coefficient for the rare-metal thermocouples are given on page 27. Clearly, the coefficients for types R and S are similar—rising from about 5 μV K^{-1} at 0°C to at least 10 μV K^{-1} at 400°C and above. The type B thermocouple is interesting in that its coefficient is similar to that of the others above 1000°C, yet falls rapidly to zero at lower temperatures. This feature arises because the Seebeck coefficients of Pt 6Rh and Pt 30Rh are similar near 0°C and their difference increases rapidly with temperature, as can be seen in Figure 2.3. Indeed, the relative Seebeck coefficient is roughly proportional to temperature (in °C) from 0 to 600°C, being numerically equal to ~1% of the temperature. The type B thermocouple produces ≤3 μV in emf for tip temperatures from 0 to 50°C for a CJ temperature of 0°C. So, if the CJ temperature is known to lie between 0 and 50°C, and an error of this magnitude is acceptable—equivalent to 1°C at 300°C and 0.3°C near 1000°C— the thermocouple may be left uncompensated (see section 3.4). In other words, its cold-junction ends may be connected to a pair of copper leads and their

(CJ) temperature ignored, after establishing that it is less than ~50°C.

It is recommended that a wire diameter of 0.5 mm be used for bare-wire Pt-based thermocouples. It gives the best compromise between stability, cost and ease of handling. A reduction in wire diameter to 0.15 mm would reduce the wire cost by an order of magnitude but increase its vulnerability to contamination effects by a factor of 3.

Rare-metal thermocouples are expensive and need protection from contamination. As a consequence, it is common practice to insert an 'extension lead' between the protected thermocouple and the instrument. The use of such leads is dealt with on page 86.

2.5.1 Effects in bare-wires up to 1200°C

Optimum conditions for Pt-based thermocouples occur when they are assembled in clean components, including a single length of high-purity, recrystallised, twin-bore alumina, with the wires exposed to air and with no avoidable sources of contamination present. Under these conditions, and with temperature limited to a maximum of ~ 1200°C, the thermocouples experience small changes in Seebeck coefficient from the following causes:
- the initial anneals if any,
- cold work arising from handling the wires,
- quenching effects on withdrawing them from a furnace and
- the oxidation of Rh around 800°C.

All data given in this section refer to a wire diameter of 0.5 mm and to thermoelements under the optimum conditions mentioned above—in alumina and under clean oxidising conditions.

Let us examine these four effects in more detail. At temperatures above 1000°C changes occur through grain growth, the relief of cold work and to other changes in microstructure. These changes tend to plateau with time and this allows the use of annealing techniques to stabilise Pt-based thermocouples before use. In one study [35] of type S thermocouples the effect of various anneals, at 1150, 1300 and 1450°C, were assessed. Relative to the as-received condition, changes of up to $5.5\,\mu$V were noted at a calibration temperature of 232°C and $4\,\mu$V at 1083°C. For comment on the use of stabilising anneals, see section 4.3.2.

Cold-work affects the Seebeck coefficient; and significant changes can occur on handling the wires, especially while installing them in twin-bore insulation. For example, running the wires over the thumb nail to remove kinks may change the coefficient by $-40\,\text{nV}\,\text{K}^{-1}$ at 0°C and $-2.5\,\text{nV}\,\text{K}^{-1}$ at 1000°C, the net result being a change in emf at 1000°C of $-16\,\mu$V, or -1.4°C [36]. Much

2.5. TYPES B, R AND S

of the cold work can be removed by a 1 h anneal at 1100°C or above, but some residual effect remains in the PtRh leg.

Various lattice characteristics, such as the order/disorder arrangement of dissimilar atoms, lattice vacancies and other defects, establish their equilibrium levels almost instantaneously at high temperatures, above 700°C say. On removing the thermocouple from temperature, and allowing it to cool rapidly to ambient, many of these high-temperature features are 'trapped-in'. The Seebeck coefficient is then temporarily low. When next used at ~ 250 to ~ 500°C, where atomic mobilities are high enough, the lattice relaxes back to a more appropriate state, lattice defects become fewer, short-range ordering of the Pt and Rh atoms may occur and the Seebeck coefficient increases. This is shown in Figure 2.5. The processes are reversible and if the wires are cycled in temperature, hysteresis will be evident—a phenomenon discussed on page 46.

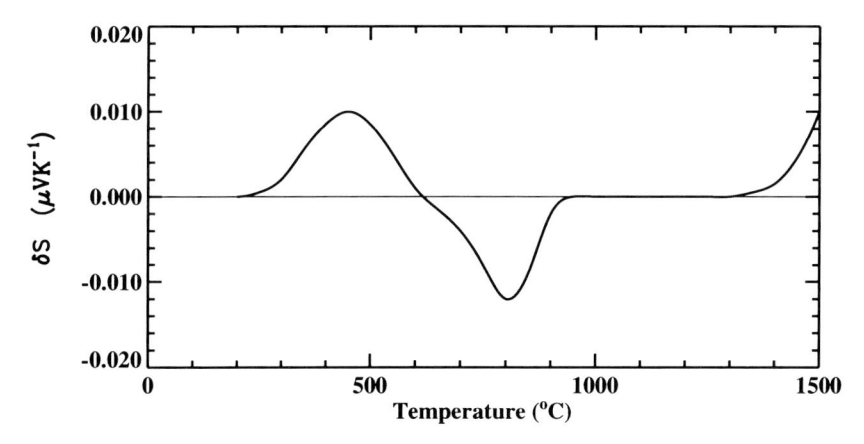

Figure 2.5: Change in Seebeck coefficient, δS, that occurs in 200 h as a function of temperature along type R or type S thermocouples in clean air. The data (calculated from data in references [37] and [38]) are approximate and relate to wires of diameter 0.5 mm, initially annealed (page 122) then quenched.

Rhodium oxidises above about 500°C and the oxide, thought to be Rh_2O_3 [35], dissociates above 900°C. As a result, the oxidation of Rh in PtRh alloys causes a reduction in Seebeck coefficient when at 500 to 900°C, with the greatest effect occurring around 800°C (the negative peak in Figure 2.5). At 800°C the Seebeck coefficient of Pt 10Rh falls rapidly with time to begin with and then slows to a linear rate from ~ 30 to at least 200 h [38], with the change at 100 h being approximately -8 nV K^{-1}. On heating to 1000°C the oxide dissociates within a minute or so and the Seebeck coefficient recovers. The formation of Rh oxide is thus a reversible process and may cause hysteresis. To illustrate the magnitude of the effect, consider a type S thermocouple, initially free of oxide, monitoring a temperature near 1000 °C for 100 h. In that region of the alloy wire from 500 to 900 °C, the Seebeck coefficient will decrease, owing to the loss of Rh to its oxide, and the net emf of the thermocouple will decrease 1.6 μV, equivalent to 0.2 °C. This was calculated as half the product of 400 °C, the range over which oxidation occurs, and -8 nV K^{-1}, the change in Seebeck coefficient at 800 °C.

For a bigger effect consider a thermocouple previously used for 100 h at 800 °C and then used for the measurement of temperature at or above 1000 °C. If the latter measurement is at a reduced immersion depth, so that the temperature gradient zone lies within that section previously at 800 °C, its emf will be low by the equivalent of 1 °C [37].

Working standard type R or S thermocouples are calibrated in either the quenched state or after given a uniform anneal at 450 °C for about 16 h (see page 119). The main cause of *in-situ* drift in pre-annealed (450 °C) Pt-based thermocouples operating at 900 to 1300 °C (tip temperature) is the oxidation of Rh. Taking the growth in Seebeck coefficient of type R and type S thermocouples at 800 °C as -0.05 nV K^{-1}h^{-1} [38] and assuming the linear rate continues indefinitely, the contribution from oxidation to *in-situ* drift would be -10 nV h^{-1}, for tip temperatures above 900 °C. This is roughly -1 °C per 1000 h, although the rate can be expected to decrease beyond 1000 h.

Type R and type S thermocouples in the quenched state will show a smaller *in-situ* drift, because the effect of oxidation is partly offset by the increase in Seebeck coefficient occurring in that section of thermocouple at 250 to 500 °C. For this reason standard thermocouples are calibrated at the CSIRO National Measurement Laboratory (NML) after a suitable quench (page 122). For further discussion on how thermocouples behave when quenched or annealed see section 4.3.

2.5.2 Effects in bare wires beyond 1200 °C

Let us now discuss effects that become significant at temperatures above about 1200 °C (depending on the quality of insulation—see below). There are two, both affecting Pt more than PtRh, unlike the changes described above:
• the vapour transfer of metal between the wires and
• contamination from the alumina insulation.

Vapour transfer of Rh, especially as the oxide, will occur whenever the wires are not fully isolated from each other by high density twin bore insulation. It will occur near the tip and at any breaks or cracks in the insulation. The magnitude of the effect depends on the lengths of exposed wire, on the duration and temperature of exposure and on the temperature gradient at the

2.5. TYPES B, R AND S

affected parts during the measurement in question. For example, a 2 mm long gap in the twin-bore insulation, positioned over type S wires at 1600 °C for 24 hours, will result in a localised transfer of rhodium and a drop of maybe 20% in the Seebeck coefficient for the exposed section of thermocouple [38]. If this 2 mm long section is then located in a steep temperature gradient region it may span ~ 20 °C say, and an error of -4 °C would result. It follows that all sections of type R or S thermocouple wire likely to experience temperatures above 1100 °C should be contained within a single continuous length of twin-bore insulation. The vapour transfer that will occur near the exposed tip of a thermocouple is limited to such a small distance that its effect is unlikely to be significant unless the tip lies in region of large temperature gradient.

Given that the twin-bore insulation is arranged to prevent vapour interchange and that it affords adequate electrical insulation between the wires there remains the question of whether it contaminates the wires. Indeed, it is the insulation, and not the wires *per se*, that sets the performance limits for Pt-based thermocouples above 1200 °C. For this reason only high-purity alumina with $< 0.04\%$ Fe should be used, such as Degussit AL23, Haldenwanger Alsint 99.7 and extrusions from Ceramic Oxide Fabricators, Vic., Australia.

Even in the best available alumina insulation the effect is significant [38]. A thermocouple fixed in position, monitoring a temperature near 1700 °C, for example, would develop an error of roughly -2 °C in 100 hours from this cause. If it is then used at shorter immersion, with most of the temperature gradient region within that section affected by the use at 1700 °C, an error of -1% would result (e.g., 10 °C loss at 1000 °C). The effect of temperatures lower than 1700 °C is less by a factor of 2.5 for each 100 °C. For example, if that section of wire given 100 hours at 1500 °C then supplies most of the emf in its next use, the error would be -0.15%.

For a single previously-unused insulator the decrease in net Seebeck coefficient is rapid at first and then levels off in about 100 h, as the source of contamination is depleted. Hence, greater changes will occur if the insulator is replaced by a fresh one, and the Seebeck coefficient could eventually level off, after repeated replacements for example, with a decrease equal to three times that suggested above [38]. On the other hand, since most of the above change occurs in the pure Pt thermoelement, the type B thermocouple is expected to be far less affected.

Studies using various grades of alumina have demonstrated the importance of using only the highest purity recrystallised insulation [39, 40] and that the type B thermocouple is less sensitive to such effects.

2.5.3 Long-term drift data for bare-wire thermocouples

Pt-based thermocouples are often used to monitor or control furnace temperatures, i.e., in long-term applications at fixed immersion. All the above-mentioned processes, where relevant, combine in the gradient region to give the reported *in-situ* drift results seen in Table 2.7. The greater drifts observed in argon and vacuum are attributed to impurities from the ceramic, which in air convert to relatively harmless oxides. The data in the table refer to 0.5 mm wires in alumina and under clean conditions.

Table 2.7: Long-term *in-situ* drift data for 0.5 mm bare-wire Pt-based thermocouples in high-purity alumina, twin-bore insulation.

Atmosphere	Type	Temperature (°C)	Time (h)	Drift (°C)	Ref.
Air	R	1450	1 000	−3	[34]
		1330	10 000	−5	[41]
Argon	R	1330	10 000	−23	[41]
	B	1330	10 000	−13	[41]
Vacuum	R	1330	10 000	−23*	[41]
	B	1330	10 000	−26*	[41]

∗ values extrapolated by authors [41], since thermocouples failed early.
Note: data in study [41] showed an inverse dependence on wire diameter, suggesting surface phenomena.

Under less than ideal conditions somewhat different behaviour may occur. For example, the effect of not using a single length of alumina twin-bore insulation may be large, even in clean air. In one study [42], a −40°C drift was noted for a type R thermocouple clad in 75 mm lengths of twin-bore while monitoring a temperature near 1400°C for 2400 hours. Much of this drift was attributed to vapour transfer of Rh in the insulator gaps. In another example [34], the consequences of not using clean air may be seen. A decrease in apparent temperature of 95°C in 500 h at 1450°C was reported for a type R thermocouple in an alumina twin-bore, and the drift for magnesia insulation, being less prone to reduction, was only −7°C. In both cases, the thermocouples were surrounded by dense alumina sheaths, to protect them from the effects of a cracked ammonia atmosphere, which tends to reduce ceramic oxides at high temperatures. Without the sheathing, the drift would have been far greater.

2.5.4 MIMS probes having type B, R or S thermoelements

Pt-based thermocouples in the MIMS configuration are unstable if a base-metal sheath is used. A 3 mm-diameter Inconel-sheathed type R thermocouple probe was assessed at NML in 1976. It was calibrated at 1000°C, uniformly annealed at 1000°C for 20 h and re-calibrated at the same depth of immersion. The brief anneal had changed the calibration by the equivalent of −11°C. The thermocouple was then scanned at 1000°C, i.e., calibrated as a function of immersion depth, to reveal a variation along the probe equivalent to 14°C in apparent temperature. Studies with 0.5 mm-diameter MIMS probes showed even greater instabilities [43]—the calibrations of type S probes at 1300°C were monitored for 20 min and the drifts were −18°C h^{-1} when in 304 stainless steel sheaths and −10°C h^{-1} for Inconel 600 sheaths. Type B thermocouples in the same MIMS configurations drifted only 40% of this amount. For the thermocouple in the stainless steel sheath the Seebeck coefficient near the tip was low by 63% and the wires had picked up Ni, Cr, Mn and Fe from the sheath.

The use of PtRh as a sheathing alloy for the MIMS probe gives some improvement, although the success depends on the porosity of the MgO ceramic and on whether Pt is present as a thermoelement. At 1427°C the *in-situ* drift of 1.4 mm diameter probes after 500 h was between −50 and −280°C, with six thermocouples of each type (S and B) giving results scattered over this range [44]. The results are related to the porosity of the ceramic and attributed to Rh vapour diffusion from the sheath. In another study, 1.5 mm type R probes sheathed in Pt 10Rh at 1450°C drifted −200°C in 1400 h [42]. These had a porous ceramic. Another specimen, especially prepared with its MgO having been sintered in use to a dense, impervious mass, drifted only −25°C in the same time. Further, it was noted from spectrographic analysis that little Rh had moved from that section of sheath whose temperature was below 1200°C. It follows, then, that PtRh-sheathed MIMS thermocouples having Pt-based thermoelements will be little affected by Rh transfer if their use is limited to temperatures below about 1200°C.

2.6 Nickel-based thermocouples (types K and N)

The first pair of type K alloys was introduced in 1906 by A. L. Marsh, a founder of Hoskins Manufacturing Co., and registered under the trademarks Chromel and Alumel. Since then, other manufacturers have produced similar alloys under the trade names, T1 versus T2, Tophel versus Nial, Isotherm-plus versus Isotherm-minus, Vacoplus versus Vacominus, etc. With the advent of universal reference tables, the letters 'KP' and 'KN' were introduced as a means of specifying the two alloy categories, each defined, not by composition, but in terms of compliance with the reference tables. The terms Chromel and Alumel, through common usage, are also used as the generic terms for these alloys. Notice that 'Chromel' is often used interchangeably with 'Chromel-P' when discussing thermocouple alloys, yet it has a more general meaning. It is used by Hoskins Manufacturing Co for their other NiCrSi alloys as well.

The alloys are similar in appearance, but Alumel is distinctive at room temperatures as it is ferromagnetic (attracted by a magnet). The reference tables for type K thermocouples cover the range −270 to 1370°C and the net Seebeck coefficient shows little dependence on temperature above 0°C. It is 41 μV K^{-1} to within ±5% over the range 0 to 1000°C (see page 27).

The early history of type K alloys is interesting [45, 46]. It was realised that the alloys, especially 'conventional' Alumel, have a limited life when exposed to air at temperatures above about 1000°C. They fail mechanically, through a decrease in diameter, and they exhibit considerable thermoelectric instability [47]. Chromel also suffers from 'green rot', the preferential oxidation of Cr in atmospheres with low but not negligible oxygen content, for example, within sheaths where air is limited and stagnant. In the period 1950 to 1962, manufacturers developed alternatives to Alumel, essentially Ni∼3Si, for improved oxidation resistance, and proposed additions of Nb, Cu and or B to the positive leg as a means of controlling green rot. It was shown [48, 49, 50, 51] that these non-conventional type K thermocouples were indeed superior to the standard formulations. For example, the loss of core diameter in Ni 3Si at 950 to 1000°C was five times less than in conventional Alumel over the same period and its thermoelectric drift was reduced by a factor of four. Despite these improvements, the new alloys, such as the Hoskins' combination 3G–345 versus 3G–196, never superseded the conventional type K thermocouple: Ni 9.5Cr 0.5Si versus Ni 5(Si Mn Al). I gather this is due to the difficulty of adjusting the 'improved' alloys so that their emf's closely matched the values given in the type K reference tables at temperatures below about 200°C.

There is also the problem of 'ageing' in type K thermocouples, which became a topic of study in the sixties [52, 53]. The Seebeck coefficient of wire at ∼ 250 to ∼ 550°C, and especially near 400°C, will increase with time; yet it will return to its former value if taken briefly to 700°C or above. This reversible, hysteresis phenomenon (discussed on page 46) occurs more in Chromel than in Alumel. In Chromel, hysteresis is related to the Cr content and was once thought to be caused by short-range ordering of the NiCr lattice [54]. A more recent suggestion by Pollock [55] that hysteresis may be explained by an electron spin-cluster mechanism appears more likely.

Quenched samples of Chromel (Ni∼10Cr) will increase in Seebeck coefficient when held at temperatures in the hysteresis zone and Ni 20Cr experiences

2.6. TYPES K AND N

a decrease in coefficient under the same conditions. So, attempts were made to find that level of Cr at which hysteresis becomes zero, with two investigators [56, 57] suggesting Ni 13Cr and one [58], Ni 15Cr. Consequently, the thermocouple Ni 13Cr versus Ni 3Si was proposed [59] as an improvement over type K. Also, investigators at Driver Harris had shown [45] that Ni 14Cr has far less hysteresis and a greater oxidation resistance than Chromel, and that if combined with Ni 3Si a superior thermocouple would result. Similar studies were conducted at Hoskins Manufacturing Co [60]. Most attempts to find an optimum thermocouple combination in the region Ni 13-15Cr versus Ni 3-4Si were considered unsuccessful/impractical because the various combinations failed to comply with the type K emf specification.

Nevertheless, Burley and his colleagues at the Australian Department of Defence, Materials Research Laboratories and Starr and Wang [61] of Wilber B. Driver, USA, continued with the development. The positive leg, now referred to as Nicrosil, began as Ni 10Cr 2Si [62], then became Ni 15Cr 1.5Si [63] and finally Ni 14.2Cr 1.4Si [64], although at least one manufacturer includes 0.1%Mg [65] (see also page 57). Similarly, the negative leg, Nisil, progressed from Ni 4Si, to Ni 4.3Si 0.13Mg and then to Ni 4.4Si 0.1Mg. The thermocouple combination was finally accepted internationally as a useful advance and was given the letter designation 'N' in 1983.

Much of the developmental work, since circa 1950, was aimed at increasing thermoelectric stability by minimising hysteresis and reducing the effects of oxidation. The most significant outcome was the type N thermocouple, whose benefits have been well established [66] (see section 2.6.2). Unfortunately, hysteresis is not easily avoided. In formulating Nicrosil, various NiCr alloys were assessed for hysteresis by heating them for an hour or so in the critical region ~ 250 to $\sim 550°C$. Under these conditions, hysteresis in Nicrosil is 5 to 10 times less than in Chromel [13]. Hysteresis also occurs in Nisil and Alumel (page 60), though to a lesser extent, and as a result the relative improvement for the thermocouple Nicrosil-Nisil, is not as good. As a consequence, net hysteresis in the above temperature range for the type N thermocouple is 2 to 4 times less than in the type K, for short periods of heating, say 15 h; although in the longer term, the gain seems less [67]. Furthermore, unlike Chromel, the Nicrosil alloy is susceptible to hysteresis at the higher temperatures, from 500 to 1000°C, peaking at $\sim 700°C$ [46, 67]. For further detail on hysteresis in Ni-based thermocouples see page 46.

The above developments took place using bare-wire thermocouples, especially when fully exposed to air. Thus, all comments on thermoelectric behaviour relate to the bare-wire and not to the MIMS format, except for hysteresis phenomena, which are intrinsic to the alloys. The behaviour of MIMS thermocouples at high temperatures is different and in this configura-

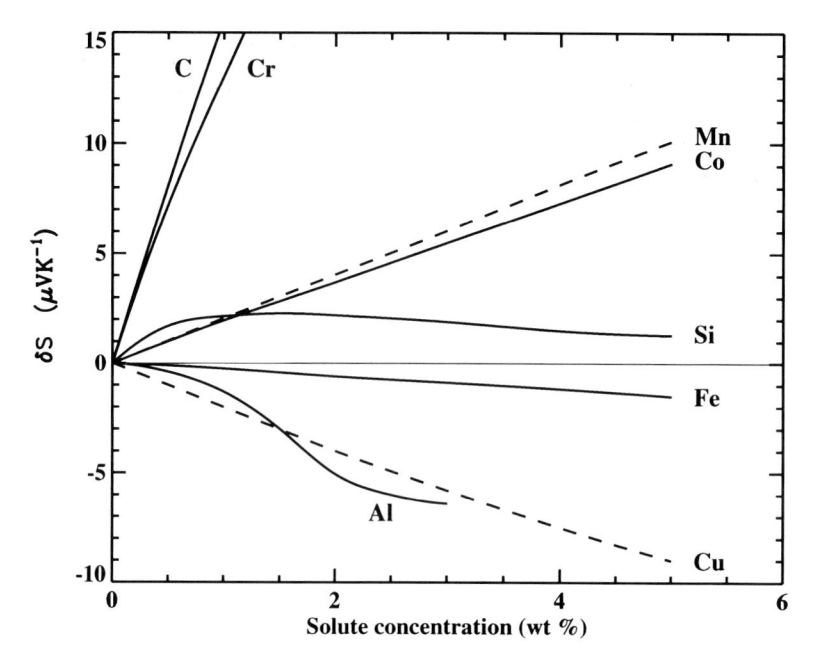

Figure 2.6: Change in Seebeck coefficient, δS, with solute concentration in binary nickel alloys, for the solutes C, Cr, ... ,Cu. Dashed lines are less reliable.

tion the relative advantage of either the type K or the type N thermocouple is not so clear (see page 67).

The reference tables for type N thermocouples cover the range -270 to $1300\,°\text{C}$ and their net Seebeck coefficient varies from $26\,\mu\text{V K}^{-1}$ at $0\,°\text{C}$ to a peak of $39\,\mu\text{V K}^{-1}$ at $740\,°\text{C}$, and then reduces to $36\,\mu\text{V K}^{-1}$ at $1300\,°\text{C}$ (see page 27).

The effect on the Seebeck coefficient of varying the solute concentration of a binary nickel-based alloy is given in Figure 2.6 for various solute elements. The data were deduced from Seebeck coefficient values [68], which contain some inconsistencies, and some emf data [13, 69] given relative to an arbitrary zero. From the figures, the effect of varying or of adding various elements to Alumel or Nisil can be estimated. It is not so easy for the positive legs because of the high Cr levels. For example, adding Si to the alloy Ni 14.5Cr changes its Seebeck coefficient by $-2.5\,\mu\text{V K}^{-1}$ for each percent Si added [13]. This is similar in magnitude, but opposite in sign, to the effect in Ni, and thus in Alumel or Nisil (Figure 2.6).

The change, δS, given in the figures will apply reasonably well to all temperatures. In other words, from equation (1.3), the change in emf of a

2.6. TYPES K AND N

thermocouple, affected by a change in solute level, will be the product of its tip temperature (assuming a CJ at 0°C) and δS. That δS is independent of T was assumed when treating the emf data [13, 69] for the figures and, within the experimental scatter in [68], the assumption is reasonable. Nevertheless, the relationship between δS and T is not this simple. The data from one study of Ni-based alloys [70] are consistent with the above assumption but those in another [71], involving changes in the levels of Mn and Al, suggest that the dependence of δS on temperature and time is rather complex.

2.6.1 Thermoelectric stability of Nickel-based thermocouples

In any thermocouple, the net Seebeck coefficient changes with time because of contributions from a variety of processes in both wires. Some of these processes are reversible and cause hysteresis (page 46) in thermocouple calibration, if the temperature is cycled, as well as contributing to drift. Reversible change and the resultant hysteresis effect occur equally in the bare-wire and the MIMS configurations. The physical and chemical changes that occur in thermocouple alloys, both reversible and irreversible, are temperature dependent and the net change in Seebeck coefficient at any one point along a thermocouple will differ from the change at another point if the temperature is different.

For example, the changes taking place along type K thermocouples are given in Figure 2.7, as a function of temperature. Here, the data indicate the net changes occuring in wires held at the indicated temperature for 200 h. The data for ID-MIMS probes (defined on page 56) were measured directly, that for bare-wires were calculated from the drift behaviour shown in Figure 2.8 and that for MIMS cable, from similar drift data. Equivalent curves may be constructed for the type N thermocouple, e.g., the reversible component (applying equally to the bare-wire and MIMS formats) may be seen in the figure on page 60.

Each curve in Figure 2.7 is equivalent to a plot of the changes taking place along a thermocouple during a fixed immersion application, plotted not as a function of position but of temperature. Consequently, it applies equally to any temperature profile along the thermoelements. If the tip temperature is less than 1200°C (extent of the plot) then consider the data only up to the tip temperature. For example, at those points (of a new type K thermocouple) having a temperature between 400 and 450°C the Seebeck coefficient would change by $0.45\,\mu\text{V}\,\text{K}^{-1}$ in 200 h, an increase of about 1%. The contribution to the total thermocouple emf from such sections of the probe is an amount that drifts upward during the initial 200 h period of use. Similarly, a region at about 1100°C in the MIMS probe of Figure 2.7 would contribute an emf component that decreases about 2%.

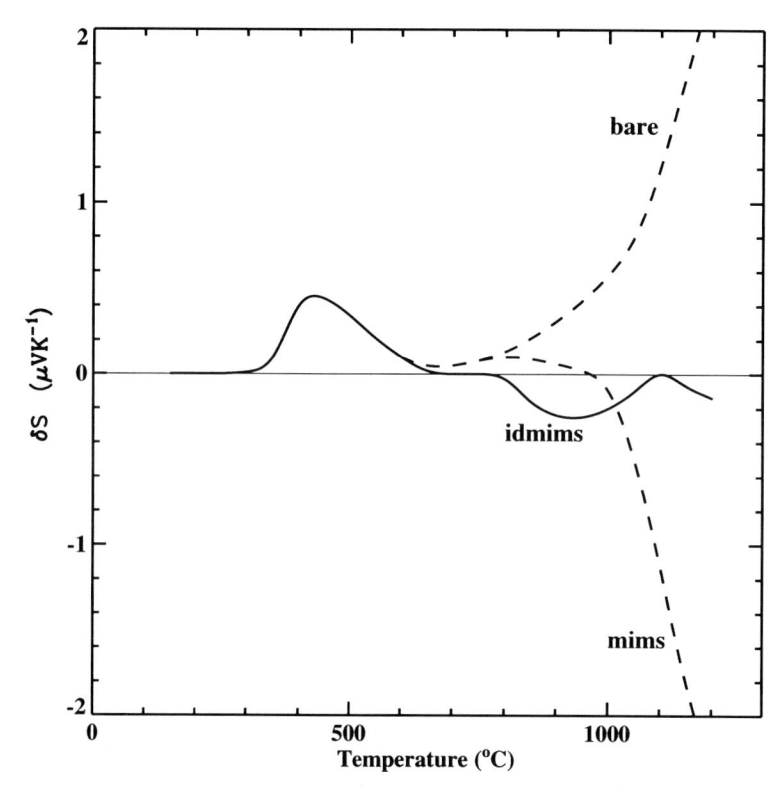

Figure 2.7: Change in Seebeck coefficient, δS, that occurs in 200 h at the indicated temperature along as-received type K thermocouples in clean air. Shown are data for bare-wire (bare) thermocouples of 3.3 mm wire diameter, the ID-MIMS probes (idmims) of Figure 2.11 and 6 mm OD SS-sheathed MIMS probes (mims). The equivalent plot for type R or S thermocouples appears on page 37.

Curves, such as those of Figure 2.7, are useful in estimating *in-situ* drift at any temperature and the effect on thermocouple emf of a change in immersion. The *in-situ* drift experienced for a 200 h period is simply the area under the relevant curve up to the temperature of the tip. An example of such calculations for the data of Figure 2.11 begins on page 62.

Reversible changes in Seebeck coefficient (hysteresis)

Up to about 600°C the only contribution to the changes evident in Figure 2.7 are those from reversible processes, a possible explanation for which is given on page 42. This is also true for type N, although in such thermocouples the dominant region for reversible change extends from $\sim$ 450 to 1000°C, peaking

2.6. TYPES K AND N

at around 650°C (see Figure 2.9). Reversible thermoelectric change, being a consequence of bulk, inherent properties of the alloys, do not depend on thermocouple format (bare-wire, MIMS etc.) nor on wire diameter. Thus the more detailed discussion beginning on page 60, based on data for ID-MIMS probes, applies in general.

Since no significant irreversible changes in Seebeck coefficient occur below about 600°C (Figure 2.7), the *in situ* drift in as-received Ni-based thermocouples for temperatures below this value will be the same for all formats. For example, a type K thermocouple will drift the equivalent of less than 2°C while monitoring a temperature near 500°C for 1000 h (Figure 2.12) and a type N less than −1°C. Such drift, and variations in calibration occurring on changing the immersion, may be greatly reduced by exploiting two properties of reversible change.

1. The change in Seebeck coefficient at any one temperature tends to plateau out (see Figure 2.10 on page 61)—with 80% of the change in 200 h occurring in the first 16 h (over night).

2. Once a change has occurred at any one temperature, a subsequent use at a lower temperature will not reverse the change—at most, it will merely add a little to the change.

Consequently, if type K thermocouples are to be used at temperatures never exceeding 450°C, say, they may be stabilised by a uniform anneal at 450°C. One such application is the repeated use of flexible test thermocouples for measuring the temperature distribution within ovens (page 203). Then, all that length of each thermocouple that is likely to be taken above ambient temperature in use could be loosely coiled in an oven and annealed over night at 450°C. The anneal would increase the calibration (emf at any temperature to 450°C) by about 1% and subsequent use will have relatively little effect. The thermocouple will remain relatively homogeneous and its calibration will be stable (provided care had been taken during the anneal to avoid the effects of binders etc., mentioned on page 178). Notice, from Figure 2.10, that the longer the period of anneal the less will be the effect of a subsequent use on either the homogeneity or the calibration. However, the anneal would need to be extended from 16 h to 3 days to improve the stability by a factor of two.

The above comments also apply in principle to the type N thermocouple, although the reversible change below 600°C is significantly less than in type K (Figure 2.7). Thus the benefits of an anneal at 450°C are proportionally less.

If the temperature of any part of a type K or type N thermocouple is taken to a higher temperature **hysteresis** in emf may be observed. It arises because of a third property of reversible change:

3. Once reversible change in Seebeck coefficient has developed in a material it is undone by a brief heating at temperatures beyond that region where such change occurs, i.e., beyond about 700 °C in type K (Figure 2.7) and 1000 °C in type N (Figure 2.9).

For example, if part of a type K probe, initially in the quenched state, is held at 450 °C its Seebeck coefficient will increase exponentially with time as indicated by Figure 2.10 and in 200 h will have changed by $0.45\,\mu\mathrm{V\,K^{-1}}$ (Figure 2.7). Then, if it is taken to 700 °C, or above, its coefficient will rapidly revert to its former value (within a minute) and will remain at this value if cooled quickly to below ~200 °C. A second use at 450 °C will again see an increase in the Seebeck coefficient as before, and so on.

Irreversible changes in Seebeck coefficient

Irreversible change occurs in those sections of the thermocouple at temperatures above about 600 °C (Figure 2.7). Such effects cause an increase in signal for bare-wire thermocouples, a change that increases rapidly with temperature. For MIMS and ID-MIMS probes of type K the change is positive for temperatures from 600 to ~ 900 °C and negative at higher temperatures. In the bare-wire case, the *in-situ* drift for type K thermocouples would be reduced if the component from reversible change is removed by a uniform pre-anneal at 450 °C. However, a reduction in drift by 2 °C in 1000 h may not be considered worth the effort for use above 1000 °C, where the likely drift is much greater than 2 °C (Figure 2.8).

On the other hand, there is no benefit in giving a MIMS or ID-MIMS probe a 450 °C anneal prior to a high-temperature use. Indeed, it may worsen drift behaviour in type K. Reversible changes in Seebeck coefficient taking place in that section of a type K probe at temperatures in the range 300 to 600 °C contribute a component of emf that increases with time and partially offsets the decreasing contribution from sections above about 900 °C (see Figure 2.7).

2.6.2 Bare-wire nickel-based thermocouples

The *in-situ* drift, defined as the drift in signal for a thermocouple with a fixed temperature profile, of bare-wire thermocouples for either the type N or the conventional type K formulations is shown in Figure 2.8. Clearly, the bare-wire type N thermocouple drifts significantly less than the type K and the improvement is due partly to the absence of Mn and Al in the negative leg.

The changes that occur at high temperatures, above about 800 °C, are due primarily to processes at the surface of each wire and are critically dependent on the presence and mix of key elements, such as Mn, Al, Si and Mg. Indeed,

2.6. TYPES K AND N

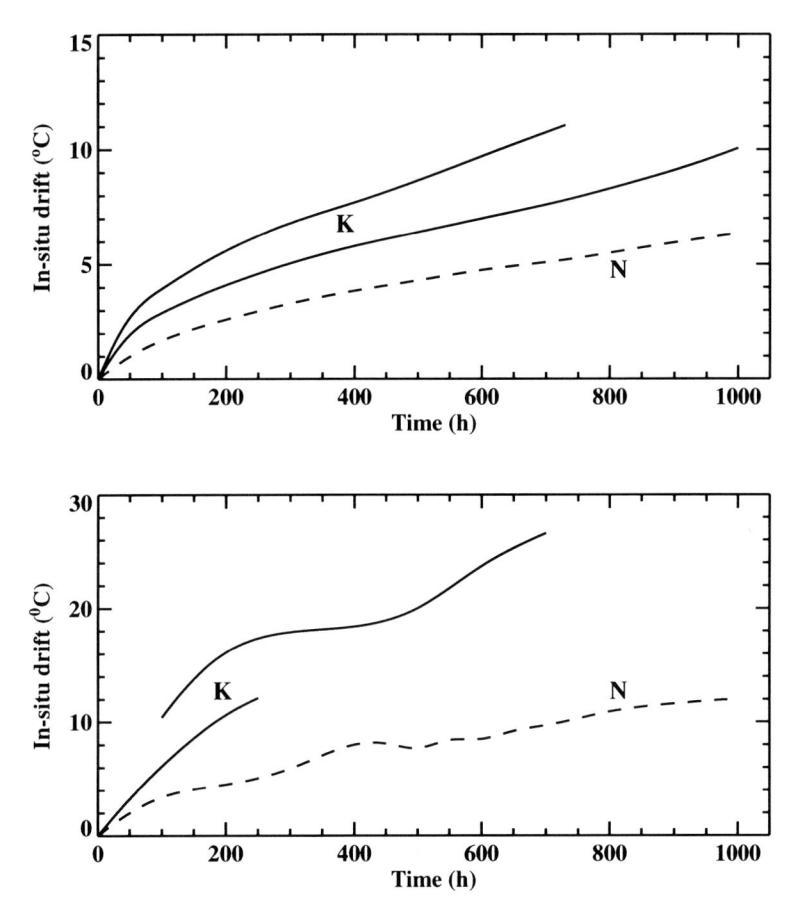

Figure 2.8: *In situ* drift, ΔT, as a function of time at 1100°C (upper) and 1200°C (lower), for type N (- - -) and conventional type K (—) bare-wire thermocouples. The lower curve for type K at 1100°C [72] relates to a wire diameter of 1.6 mm, as do the type N data [72], and the others to 3.3 mm [63, 64].

a 5 to 1 variation in drift rates was observed in one study [72] of type N thermocouples and the scatter was attributed mainly to the ~0.1% range in Mg content of the wires. The slight variation in composition arose because the alloys were sourced from different manufacturers.

Likewise, drift rates will differ if the supply of O_2 is limited, if gases other than air are present or if the wires are contaminated by carbon or sulphur. Consequently, the drift behaviour of bare-wire thermocouples at high temperatures is not repeatable and cannot be predicted from laboratory results like those given in Figure 2.8. Here, data refer only to clean, oxidising conditions and are sample dependent (small variations in composition may have a significant affect on behaviour). Larger drifts are likely in many industrial situations and, in practice, thermocouple drift will depend on how the wires sit within their insulators and, thus, on their access to oxygen and other gases and contaminants.

Type K and type N thermocouples do not perform as well in a vacuum (where preferential volatilisation of Cr occurs), in a reducing atmosphere or even in a marginally reducing atmosphere. If conditions inadvertently become slightly reducing or involve stagnant air the type N thermocouple is affected less than the bare-wire type K [73]. When reducing conditions are likely, however, it is better to use a MIMS probe with a separate protective sheath, if necessary, or, if temperatures do not exceed 750°C, and a bare-wire variety is preferred, a change to the type J thermocouple (section 2.7) is preferable.

Most of the available data on *in-situ* drift of bare-wire thermocouples are summarised in Table 2.8 and refer only to ideal conditions (in clean air). In general, the drift rates for type N thermocouples are less than those for type K and the non-conventional type K alloys (K* in Table 2.8) are intermediate in behaviour. In any case, it is difficult to judge one thermocouple type against another from such data or to predict behaviour in any given application, for the reasons given above. Moreover, the tabulated values are net changes and small drift rates may be deceptive—they are not necessarily a consequence of stable Seebeck coefficients. Often, the signal from one wire drifts in the same direction as the other and, as a result, the net change is relatively small.

At high temperatures the type N thermocouple in bare-wire form has an enhanced survival rate—its mechanical integrity is retained for longer periods at temperatures above 1000°C. For example, conventional type K wires, of 3.3 mm diameter, completely oxidise in about 300 h at 1200°C [47], yet type N wires of less than half the diameter (1.5 mm) survive at least 2500 h [74]. Even at 1000°C, the loss of Alumel metal through oxidation is significant, maybe 40% in 1000 h for a 3.3 mm wire [47]. Other examples of better survival in type N may be seen in Table 2.8.

Some thermocouple applications require a small wire diameter and involve changes in the immersion depth, and as a consequence are characterised by relatively large changes in calibration. For example, the calibration of type N thermocouples of 0.5 mm diameter changes by ~32°C at 1100°C if the immersion is reduced after a 16 h use at this temperature [20] and by ~80°C if moved after 64 h [75]. Type K thermocouples of the same diameter can be expected to change by a similar amount [20]. As a general rule, errors resulting from a shift in immersion at high temperatures tend to be an order of magnitude greater than the drift that had occurred in the previous use at fixed immersion. For example, the above change of 32°C had occurred on

2.6. TYPES K AND N

Table 2.8: *In situ* drift data for bare-wire type N and type K thermocouples held in clean air for various times at selected temperatures. Included are data for non-conventional type K alloys (K*, i.e., those not containing Mn and Al).

Wire Diameter (mm)	Temperature (°C)	Time (h)	Drift (°C) K	K*	N	Ref.
0.5	500	10	1.1		−0.3	[20]
		200	1.7		−0.6	[20]
	900	10	0.4		0.5	[20]
		200	2.9		1.2	[20]
	1100	10	1.8		1.3	[20]
		200	10		3.6	[20]
	600	5000	1		−1	[74]
	800	5000	7		1.5	[74]
	1000	1000	19		5	[74]
1.5	1000	5000	20		2.5	[74]
1.6	1100	1400	8‡		1 to 9†	[72]
		3000	-		4 to 16†	[72]
	1200	400	1‡		1 to 8†	[72]
		1200	-		−3 to +14†	[72]
3.3	1000	700	3.5		1.5	[63]
		800	4		1.5	[64]*
		1000	∼5			[47]
	1100	1000	11			[47]
		700	11	−1	∼0	[63]
	1200	700	19	6	−1	[63]
		700	29			[64]*
	1250	700			6	[64]*

† range of results for 4 thermocouples from different sources.
‡ thermocouple failed at this time.
* data extrapolated to zero time.

moving a type N thermocouple after a fixed-immersion use at 1100°C, where it had drifted by < 2°C in 16 h (see also section 1.4.2). It should be noted that such drifts, and the subsequent changes in calibration on decreasing the immersion, are very dependent on temperature and wire diameter.

2.6.3 MIMS probes

MIMS thermocouple cable is manufactured usually by a drawing process. To begin with, thermoelement rods are inserted into long beads of a crushable, porous ceramic, typically MgO, and then into a metallic tube. Alternatively, the rods are held in position within the tube while MgO powder is packed in around them. The rods and tube are then swaged together at one end and the assembly is pulled through a series of dies to the required diameter of cable, which may be as small as 0.5 mm, but is more commonly 1.5 to 6 mm. After each draw the cable is usually annealed. The first pass through a die is a 'sink draw', in which both the inner and outer diameters of the sheath are reduced by similar amounts, and the ceramic is crushed to fill the available space. Subsequent draws, applied to 'solid' cable, maintain the ratio of diameters for the sheath and thermoelements.

Fabricating thermocouples from MIMS cable requires a more skilled operator (see below) than does the bare-wire thermocouple because of the welding requirements and the need to keep the hygroscopic MgO dry. Further, the choice of sheathing material is critical to the stability of MIMS thermocouples at high temperature, above 800°C say, as discussed in sections 2.5.4 (for Pt-based thermocouples) and 2.6.4. Despite these drawbacks the MIMS thermocouple is becoming more popular than the bare-wire versions because of its following advantages [76].

- The sheath is a barrier against the corrosive effects of the atmosphere, especially oxidation.

- For a similar life expectancy at high temperatures the MIMS thermocouple is smaller and less massive than the bare-wire equivalents.

- It is therefore more responsive to changes in temperature, and the conduction of heat away from its tip is less.

- It is flexible and resistant to shock.

- The sheath serves as a convenient shield against electrical interference.

- As shown in section 2.6.5, nickel-based MIMS thermocouples of good design out-perform the heavier, bare-wire equivalents in thermoelectric stability and useful life.

MIMS thermocouples are available in various tip-weld arrangements. The bonded junction (thermocouple tip and sheath welded together) offers better thermal contact with the site of interest, but the alternative, the isolated junction, is likely to be more practical. It allows the insulation resistance to

2.6. TYPES K AND N

be measured and used as a check on whether moisture had been absorbed (see below). Furthermore, the isolated junction avoids the formation of a mixed alloy at the tip weld, which may compromise survival at high temperatures.

In 1984, an X-ray study [77] was made of the tip welds of 168 probes having stainless steel or Inconel sheaths and diameters of 3 and 6 mm. The probes were from two manufacturers/fabricators. Of those from one manufacturer, 38% were judged faulty and from the other, 58%. The faults were small pinholes, cracks and regions where the weld was too thin. The problem was explained to one of the manufacturers who was then asked to re-do the welds on 27 of the probes. Even knowing their earlier failure rate and realising the welds would be assessed, 11% of the re-welds were rejected. This episode demonstrates the relative difficulty of producing quality tip-welds for MIMS probes.

If the weld is faulty the atmosphere will enter the internal space containing the thermoelements, and the survival time and thermoelectric behaviour will tend to be that of bare-wire rather than MIMS thermocouples. Performance will be poor because the thermoelements are relatively small in diameter ($1/6$ to $1/5$ of probe diameter). The problem is aggravated by the insulation—oxide impurities may be reduced and cause further damage to the wires. It is worth noting that type N thermoelements are likely to operate better in a faulty MIMS probe than are type K, because of their better bare-wire characteristics (section 2.6.2).

Fortunately, it is possible to develop a reliable welding technique, e.g., by examining the weld (X-ray and/or sectioning) after each of many attempts, as a means of adjusting the technique for each welder until success is assured. The welder gets to know what a good weld behaves and looks like.

Similarly, the difficulty and importance of forming stable hermetic seals at the 'cold' end of a MIMS probe is usually overlooked, as is the need to avoid damage to the seal during shipment, storage and use. When a seal is faulty, moisture is rapidly absorbed in the MgO insulant to reduce its electrical insulation resistance. Also, moisture will affect surface scales on the thermoelements and both effects may cause premature failure and erroneous temperature readings [78].

In general, the insulation leakage between thermoelements in a MIMS system is somewhat greater than that for bare-wire thermocouples that are loosely fitted with hard-fired ceramic beads. Insulation leakage tends to shunt some of the Seebeck emf developed closer to the tip and could have a significant effect at high temperatures, depending on immersion depth, diameter of cable, quality of insulation and the temperature profile (see item 9 on page 178).

The form of insulation used in MIMS cable poses another interesting problem—it is more difficult to arrange open-circuit-thermocouple protection.

When open-circuited, a MIMS probe at high temperatures could well present a resistance of 1 kΩ or less, which is still relatively low for modern, high-impedance instruments. In such cases, the measured temperature would correspond to a point some distance back from the tip. If the apparent temperature is within 50°C, say, of the tip temperature, it is likely to go unnoticed for some time, until measurements on the product, or its failure, reveal the discrepancy. More than one hundred nickel-based MIMS thermocouples have been studied at the National Measurement Laboratory for long-term drift over 200 to 3000 h at temperatures from 1000 to 1300°C. The only failures due to an open-circuit or short-circuit have been with stainless-steel sheaths when above 1000°C. The effects are probably related to differences in thermal expansion and to the migration of problem-elements, such as Mn, and are thus far less likely in the ID-MIMS designs (section 2.6.5).

2.6.4 MIMS probes sheathed in Inconel or stainless steel

MIMS thermocouples are available with sheaths of Inconel, 600 or 601 (see page 77), and a variety of stainless steels, such as 304, 310, 316, 321, 347 and 446 (some of which are described in section 2.10.3). For convenience, I refer to these varieties as conventional MIMS thermocouples. See page 52 for general comments on the MIMS system.

There is relatively little information published on the behaviour of conventional MIMS probes. It is known that elements migrate between thermoelements and sheath [79], that the differential thermal expansion of the sheath relative to the negative thermoelement causes failures on temperature cycling [27] and that Mn from the sheath is a major contributor to drift [70, 80].

What little data there are on *in-situ* drift, suggest similar magnitudes for stainless-steel sheathing to those of Figure 2.8, but in the opposite direction. For example, probes of type K and type N with 3 mm diameter sheaths both drift about −10°C in 1000 h at 1100°C [80] and −24°C at 1200°C [43]. Notice that 1000 h represents less than six weeks continuous operation. For Inconel sheaths the drift would be less because of its lower Mn content—at high temperatures the effect of sheath-Mn on drift tends to be proportional to the level of Mn [70], up to 1% in Inconel and 2% in stainless steel. The above data from reference [80] relate to sheaths of 310 stainless steel containing 1.5%Mn. For a further improvement it is necessary to avoid Mn in the sheath (section 2.6.5).

Drift behaviour is often erratic—in 200 h at 1100°C the drift for 10 stainless-steel-sheathed probes ranged from −1 to −8°C, and of 14 probes taken above 1100°C ten failed because of short circuits [80].

In an Inconel sheath and at the relatively low temperature of 875°C, where

2.6. TYPES K AND N

the diffusion of elements from the sheath would be less significant, type K and type N MIMS probes drift to a similar extent [65]. This result, however, depends on the source of the thermoelement material, because much of the change would arise from hysteresis (page 46). Held for 10,000 h in air, the 3.2 mm OD type K probes drifted +4°C and the type N, +3°C. In both cases, the drift beyond 5,000 h was +1°C.

In the same study [65], the drift in 300 h when in wet, cracked ammonia (reducing) was approximately +1°C, similar to that in air, but in natural gas both types drifted −5°C (300 h).

In Endogas (carburising potential from 0.3 to 1) [81], probes of 3 mm diameter, sheathed in 310 stainless steel and Inconel 600, survived at least 2500 h when held at 750 to 930°C. This study also produced some thermoelectric drift data on type K and type N probes, for both sheathing alloys. However, it is difficult to draw any definitive conclusion from the data, for the reason given in item (f) on page 68.

MIMS probes of small diameter, say 1.5 mm, are often used in situations where their temperature profile is changing, such as in conveyor-belt furnaces, or re-used at differing depths of immersion, such as in testing heat-treatment furnaces. The changes in calibration that occur on 'moving' the emf-producing zone are just as large as those occurring with bare-wire thermocouples of small wire-diameter (section 2.6.2), typically 0.5 mm, when used in such applications. For example, after a 64 h use at 1100°C, a stainless-steel sheathed type K MIMS probe changed in calibration by about −80°C on reducing its immersion [75].

When analysing or predicting the thermoelectric behaviour of MIMS thermocouples or, indeed, any thermocouple, it is best done by assessing the local changes in Seebeck coefficient that occur along the probe, as discussed in section 1.3.

The Seebeck coefficients of MIMS thermoelements undergo change from three causes. Firstly, the annealing effect of being at a temperature beyond about 600°C—this tends to undo the effects of cold work and mechanical stress. Secondly, there are reversible hysteresis effects that extend to about 1000°C and, thirdly, serious degradations in thermo-emf arise from contamination, mostly by elements migrating from the sheath [79, 80] (see also section 2.6.5). The first two are intrinsic properties occurring throughout the bulk of each metal and, thus, affect thermocouples of differing diameters to a similar extent. On the other hand, contamination has an effect that increases as the diameter decreases.

2.6.5 Integrally-designed MIMS probes

The components of a MIMS probe are usually selected on the basis of independent requirements—the need for a sheath to survive in a chosen environment and the need for thermoelements to match a particular instrument. This independent, and somewhat arbitrary, combination of materials is usually made without concern for the possible interactions and, until about 1983, there had been little published design effort aimed at minimising such effects. Moreover, many of the failures that occur in practice and the variability in performance that is seen between different probes of the same nominal type arise from poor fabrication techniques (section 2.6.3) and the quality of insulation. Therefore, what is needed is a probe that satisfies the following criteria.

- Its sheath and thermoelements are chemically compatible—to minimise instabilities in Seebeck coefficient caused by the migration of component elements. It is especially important to avoid elements of low vapour pressure, such as Mn in the sheath (see below).

- The coefficients of thermal expansion of its sheath and thermoelements are sufficiently close to cause few failures (open-circuits) on temperature cycling, see (c) below.

- It has a long life at high temperatures—a property dictated mainly by interactions at the outer surface of the sheath, e.g., oxidation and spalling.

- The ceramic insulation is of high quality.

- Its tip weld is continuous and nowhere thinner than the sheath wall (see section 2.6.3).

- At the 'cold' end it has a stable hermetic seal applied while the insulant was moisture-free (section 2.6.3).

- It has reproducible and predictable thermoelectric behaviour.

Probes having these properties, and designed as integrated systems, are referred to in this book as 'integrally-designed MIMS' probes, or **ID-MIMS**, to distinguish them from the 'conventional' probes described in section 2.6.4. In essence, considering only the first two criteria above, an ID-MIMS probe is one in which the sheath alloy is selected to minimise the effect of itself on the thermoelements.

The development of ID-MIMS probes began from the observation that the Mn content of conventional MIMS sheaths correlates well with *in-situ* drift

2.6. TYPES K AND N

data and, thus, that much of the drift would be eliminated if Mn-free sheaths were used [80]. Early in the study, in 1983, CSIRO obtained the cooperation of MM Cables Pyrotenax, Australia, in developing MIMS probes based on a sheath that is free of Mn. As a result, the first production coils of ID-MIMS cable were produced by Pyrotenax in August 1985.

The sheath alloy used, 84Ni 14.2Cr 1.4Si 0.15Mg, was referred to initially as Nicrosil, for simplicity, but was later called **Nicrosil-plus** to avoid confusion and to emphasise the important addition of Mg. The alloy was chosen for the following reasons:

(a) The long-term survival of bare-wire type N thermocouples, Nicrosil versus Nisil, had demonstrated that Nicrosil performs well as a high-temperature alloy, especially under oxidising conditions.

(b) The presence of small quantities of Mg was shown [72] to have an adverse effect on the thermoelectric stability of Nicrosil thermocouple wires, but enhances its oxidation resistance. It follows that Mg should be avoided when forming a thermoelement of Nicrosil. But, equally, it follows that in developing a Nicrosil-based sheathing alloy, where thermoelectric stability is not relevant, the presence of Mg is advantageous. Thus, 0.15% Mg was added to Nicrosil to form the new sheathing alloy, Nicrosil-plus, giving an order of magnitude improvement in oxidation resistance.

(c) The coefficient of thermal expansion of Nicrosil-plus more closely matches that of the Ni-based thermoelements than does that of any conventional MIMS-sheathing alloy (see Table 2.4). In Ni-based MIMS thermocouples it is the negative thermoelement, Alumel or Nisil, that is most likely to fracture from thermal-expansion stresses when temperature is cycled [27]. The coefficient of expansion of Nicrosil-plus differs by only 1% from that of Alumel and 5% from that of Nisil. How well the coefficients of sheath and thermoelement need to be matched depends on the particular thermoelement—it is especially critical for the type N thermoelement Nisil [27] because of its greater grain size and, thus, its propensity to fracture (see item 5 below).

Since 1983, the ID-MIMS probe and its potential has been the subject of numerous studies. The following is a summary.

1. The first step in the development of the ID-MIMS system was to verify that the avoidance of sheath-Mn gives the predicted improvement. This was done in a study [70] using type N thermoelements, which contain no Mn and thus simplify interpretation.

2. Small diameter (1.5 mm) type N ID-MIMS probes were assessed for their short-term variable immersion applications [75]. The study was conducted for an organisation interested in measuring the temperature-distributions in heat-treatment furnaces at temperatures up to 1100°C. The type N probes were found to change calibration the most when re-used at a reduced immersion. Under these worst-case conditions, the thermo-emf changed less than 0.5% after 64 h at temperatures up to 1100°C. This is a ten-fold improvement over that for 24 AWG bare-wire type N thermocouples (+6.5%) and for MIMS probes with stainless-steel sheathing (−7.5%).

3. Hysteresis phenomena in type N alloys were studied at CSIRO [67] for ID-MIMS probes with Nicrosil-plus sheaths. In such probes, measured changes in Seebeck coefficient are not swamped by the effects of contamination from the sheath or by oxidation, as would be the case if bare-wire thermocouples were studied. The study followed the initial work on ID-MIMS probes, which revealed hysteresis effects in type N alloys at a level comparable to that in type K, but over different temperature regions.

4. The majority of thermocouple users have instrumentation and cables to suit the type K emf-temperature relationship. As a consequence, CSIRO/Pyrotenax tried to develop an enhanced ID-MIMS probe by optimising the compositions of type K thermoelements. Five type K formulations, some with Mn and Al and some without, were examined. The probes had Nicrosil-plus sheaths and the study, detailed in references [82] and [83] and summarised on the following pages, included two sources of type N material for comparison. It was concluded [84] that in the ID-MIMS form, the type K and type N thermocouples behaved equally well, overall, and that the stability of the type K is to a large extent dictated by the Mn/Al content of Alumel.

5. In a study by Fawcett and Wilson [85] the use of Nicrosil instead of Inconel as the sheathing alloy did not significantly improve intergranular attack in Nisil or its loss of mechanical integrity on thermal cycling. The 3 and 6 mm diameter test probes were repeatedly cycled between ambient and 750°C. This suggests that Nisil, if it is to be used in MIMS form, needs further development.

It is useful to recognise that the type K and type N alloys were developed for use as bare-wire thermocouples in air. In other words, the alloys were designed specifically to optimise the oxidation mechanisms at their surfaces when fully exposed to air—they were not designed for the MIMS environment. Now that the more-stable ID-MIMS probe is available, the question of optimising the composition of thermocouple alloys for this confined environment could be

2.6. TYPES K AND N

considered. There is not a lot that can be done for the type N alloys because their compositions are fixed, by definition, but there is room for tinkering with the type K. One such proposal, to optimise the Mn and Al levels in Alumel, is made in reference [86].

2.6.6 Properties of ID-MIMS thermocouples

What follows is a description of the changes in Seebeck coefficient observed in the above CSIRO studies on type K and type N alloys (items 1 to 4, on page 57). Collectively, they contain most of the published data on thermoelectric change in probes known to satisfy the ID-MIMS criteria (page 56). Whereas the data refer specifically to probes sheathed in Nicrosil-plus, they should apply to any ID-MIMS probe, provided the thermoelements are similar. Furthermore, they give an insight into the thermoelectric behaviour of all Ni-based probes.

Included in the work were five type K formulations, differing in composition, especially in the levels of Mn and Al, and two sources of type N alloys with the same nominal compositions.

To ensure that any differences observed were due to the starting alloys, and not to the method of manufacture of MIMS cable, all probes were from coils produced by an identical process—that used by MM Cables Pyrotenax. Moreover, all probes were welded and hermetically sealed by Pyrotenax, who had developed techniques to overcome the little-appreciated, but common, problems associated with these procedures (discussed in section 2.6.3). As a result, the probes satisfied all the criteria for ID-MIMS probes stated on page 56.

The measurement technique was designed so that both types of change in Seebeck coefficient could be observed—the reversible (hysteresis) and the irreversible components. The reversible changes were measured relative to a reference state achieved by holding the probe at a temperature near 1000°C for a few minutes and quenching it—by withdrawing it quickly and allowing it to cool in air at ambient temperature. It produces a similar state of quench to that usually produced at the final stage of MIMS manufacture.

The changes in Seebeck coefficient that occur in Ni-based thermoelements while held at temperature, are given as a function of the temperature in Figure 2.9. The data in Figure 2.10 illustrate how the coefficient varies with time of heating in the three hysteresis regions of Figure 2.9. The type K data refer to the Mn/Al-bearing alloys referred to as K1 in reference [71] and the type N curve shows the average behaviour for the two sources of N-alloy studied (N1 and N2).

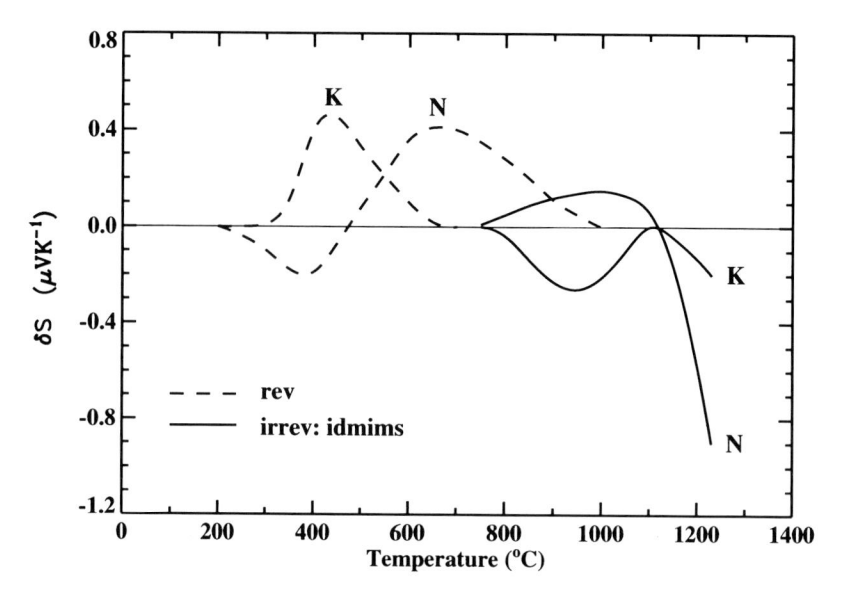

Figure 2.9: Net reversible and irreversible changes in Seebeck coefficient, δS, occurring after 200 h at temperature, for type N (mean of N1 & N2) and type K thermocouples within 6 mm diameter ID-MIMS sheaths of Nicrosil-plus. Data are from ref. [71, 87].

Reversible changes in Seebeck coefficient (ID-MIMS)

As shown in Figure 2.9, the degree of reversible change in type K and type N probes is similar (see below), being up to ~1% in 200 h for both, although it occurs in different temperature regions. Notice that the effect is not related to the thermocouples being in the ID-MIMS format—the data refer equally to conventional MIMS and to bare-wire thermocouples. In each case, the change develops rapidly to begin with and levels off after about 200 h (Figure 2.10), with maybe 80% of the change occurring in 16 h (overnight) and ~60% in 1 h (type K).

The reversible change indicated for type N probes, in Figures 2.9 and 2.10, is the average of the changes for two sources of material, referred to as N1 and N2 [83]. The change observed for N1 was twice that for N2 and so at 650°C it amounted to 1.3% in 200 h (compare Figure 2.9)—higher than the peak for type K. It should be noted, that type N alloys from manufacturers not considered in the CSIRO study may exhibit different levels of reversible change than those found. The cause of this variation among different sources of material is unknown. It is most likely due to differences in microstructure or to the levels of minor impurities, such as C—producing $Cr_{23}C_6$ [87], especially

2.6. TYPES K AND N

for the temperature zone centred on 650 °C.

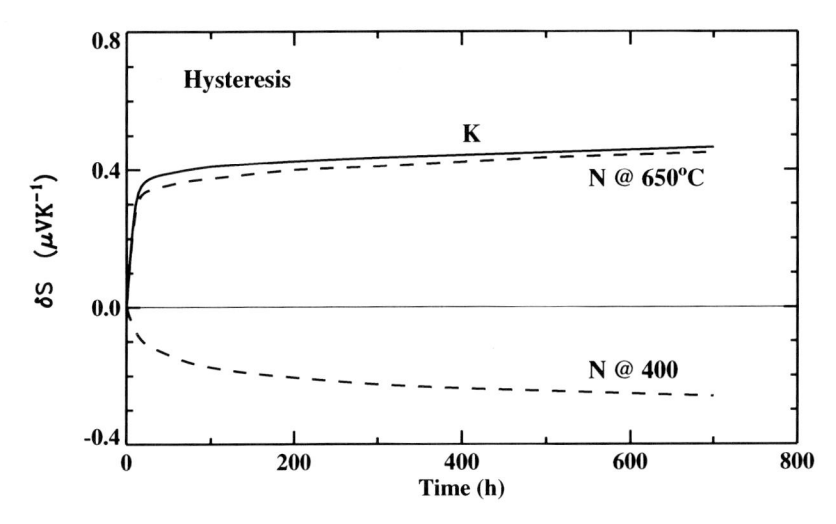

Figure 2.10: Time dependence of the hysteresis peaks in Figure 2.9, near 400 and 650 °C, for type K and type N probes.

There is one significant hysteresis peak for each of the positive thermoelements, Chromel and Nicrosil, but it occurs at different temperatures—at 400 °C for Chromel and 700 °C for Nicrosil. By contrast, the change for each of the seven negative thermoelements [87] considered is similar, in that each has two hysteresis peaks: one at 350 and one at 600 °C. Nevertheless, the changes in the Mn- and Al-bearing Alumel's are less than in the Nisil's and they would contribute little to high-temperature *in-situ* drift (>700 °C) of a type K thermocouple because their two peaks have similar areas and are of opposite sign. Notice that in Figure 2.9, only one (reversible) peak is evident for type K thermocouples. This is because it is the net effect of the two thermoelements that is plotted.

The overall effect of reversible change in the two types of thermocouple is similar. For both types major peaks of about 1% in Seebeck coefficient occur and hysteresis contributes about 2 °C to long-term drift at high temperatures. On the other hand there are significant differences:

- Hysteresis in the type K thermocouple predominates at a lower temperature than it does in type N—around 400 °C for K rather than 700 °C for N.

- Hysteresis in the type K thermocouple extends over a smaller temperature range—from 200 to 600 °C for the Mn- and Al-bearing type K alloys compared with 200 to 1000 °C for the type N.

- Hysteresis in type K alloys is reproducible and varies little from manufacturer to manufacturer, whereas in type N it is source dependent and possibly batch dependent. Alloy manufacturers, not knowing what process parameters dictate the level of hysteresis in their type N alloys, would need to assess each batch to confirm that the level of hysteresis is acceptable. Nevertheless, it is possible (and likely in the near future) to obtain batches of type N material free of significant hysteresis.

Irreversible changes in Seebeck coefficient (ID-MIMS)

The irreversible changes in coefficient shown in Figure 2.9 relate to a 6 mm diameter Nicrosil-plus sheath. For smaller diameters the changes occur more rapidly and, at temperatures above about 1000°C, are larger, tending to vary inversely with diameter. The negative thermoelements contribute the most to irreversible change, and the change is not related to whether the thermoelement is Alumel or Nisil but to whether it contains Mn and Al [71]. In Figure 2.9, the irreversible change shown for type N is due mainly to the Nisil thermoelement and is similar to the effect of Alumel on a type K probe if the Alumel contains neither Mn nor Al.

The presence of these elements in the negative leg causes two additional processes of change for the Seebeck coefficient. Firstly, the valley from 800 to 1100°C, seen in the curve marked K in Figure 2.9, is attributed to a surface loss of Al from Alumel.

Secondly, the addition of Mn to Alumel, and thereby its loss with time from the surface, causes a positive contribution to δS. This explains why the irreversible change shown by curve K in Figure 2.9 is smaller (less negative) at temperatures above 1100°C.. The opposite effect occurs when Mn-bearing sheathing alloys, like stainless steel, are used. Then, the gain in Mn by the thermoelements causes δS to become more negative than given by curve N and explains why conventional MIMS probes have a large negative drift at high temperatures (section 2.6.4).

In-situ **drift**

The combined effect of reversible and irreversible processes on the Seebeck coefficient is given as a function of temperature in Figure 2.11. Curves of this type are useful when analysing the *in-situ* drift behaviour of a thermocouple and in estimating the effect of changing its immersion. Consider a type N thermocouple whose characteristics match that shown in Figure 2.11. If used to monitor a temperature near 650°C for 200 h its drift over this period would be the integral of $\delta S(T)$ from 0 to 650°C, i.e., the area under the curve for

2.6. TYPES K AND N

type N. For that section of thermocouple from 200 to 470°C the emf would have changed by $\frac{1}{2} \times (-0.2\,\mu\text{V}°\text{C}^{-1}) \times (470\text{-}200)°\text{C} = -27\,\mu\text{V}$ (see Figure 2.11—area of a triangle is $\frac{1}{2}$ the height × the width). Similarly, the area from 470 to 650°C is 41 μV to give a total of 14 μV equivalent to 0.4°C of *in-situ* drift. If the immersion of the thermocouple is decreased so that the temperature gradient region lies within that section of thermocouple previously in the hot zone, at 650°C, the change in Seebeck coefficient for wire now in the temperature gradient region would be $\sim 0.4\,\mu\text{V K}^{-1}$. Thus, thermocouple emf would increase by $0.4\,\mu\text{V}°\text{C}^{-1} \times 650°\text{C} = 260\,\mu\text{V}$ and the temperature indication would be high by $260/39 = 6.7°\text{C}$ (Seebeck coeff. at 650°C is $39\,\mu\text{V K}^{-1}$).

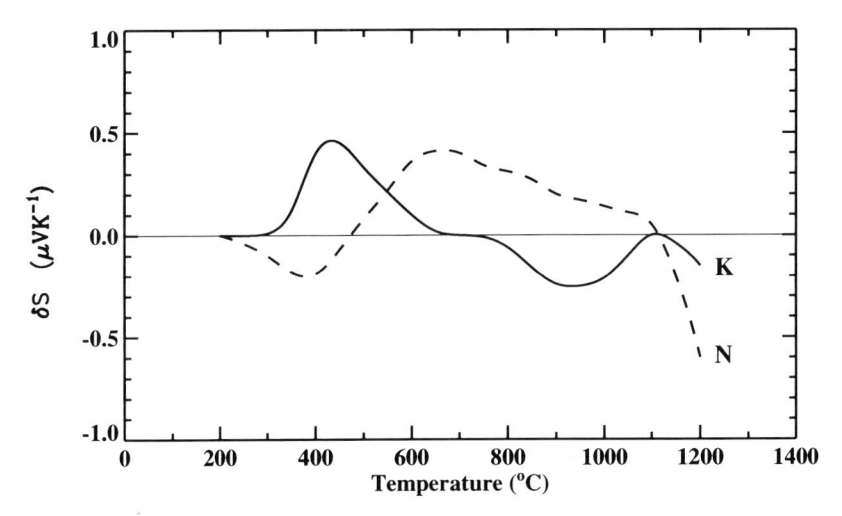

Figure 2.11: Total change in Seebeck coefficient, δS, that occurs in 200 h at temperature for type K and type N (mean of N1 & N2) thermocouples within 6 mm diameter ID-MIMS sheaths of Nicrosil-plus, from ref. [88]. Compare with figure on page 46.

In-situ drift in Mn/Al-bearing type K thermocouples of ID-MIMS design is due entirely to reversible changes in Seebeck coefficient for tip temperatures up to about 800°C. At these temperatures the emf increases by about 60 μV in 1000 h, the equivalent of 1.5°C ($S \sim 40\,\mu\text{V K}^{-1}$), and the drift is not dependent on probe diameter, *per se*. Drift beyond 1000 h will be slight because reversible change in S varies as the logarithm of heating time [87]. An example of (reversible) drift is given in Figure 2.12, where the behaviour of type K probes is compared to that of type N from the two sources.

At higher temperatures, drift behaviour is more complex because of contributions from irreversible processes and is dependent on probe diameter. The drift in Mn/Al-bearing type K probes is less than that in probes without

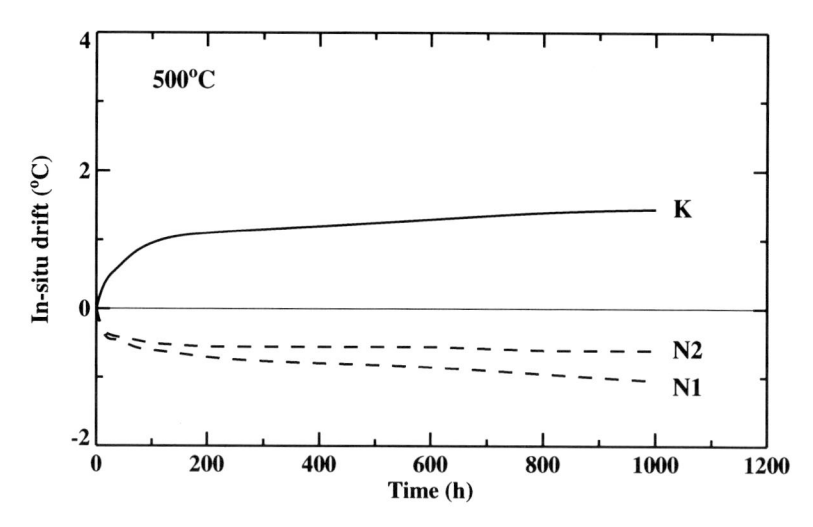

Figure 2.12: *In situ* drift as a function of time for 1.5 mm diameter ID-MIMS probes with Nicrosil-plus sheaths and tip temperatures held at 500°C, from ref. [89]. The thermoelements are type K (K) and type N (N1, N2).

Mn and Al, the type K* thermocouple. For example, at 1000°C a drift of 4°C was observed for K* over 1000 h and < 1.5°C for the Mn/Al-bearing type K (see Figure 2.13); and drift data for 1200°C are given in Figure 2.14.

Apart from thermoelectric data obtained at CSIRO, and described above, there are the *in-situ* drift data of Parker [90]. At 1100°C the changes observed in two type N probes with 1.5 mm diameter Nicrobell-A sheaths were 3 and 5.5°C in 1000 h—Nicrobell-A, a trademark of Nicrobell Pty Ltd, has the same composition as Nicrosil-plus (see section 2.6.7). At 1200°C two other probes drifted about −5°C in 350 h. For probes sheathed in Nicrobell-B (Nicrosil-plus with the addition of 3% Nb) the drift was similar, while intact (see below). These values are relatively large compared to that found in the CSIRO study, mainly because of the smaller diameter, yet they are low compared to the drift in similarly-sized conventional MIMS thermocouples. The 3% Nb then used in Nicrobell-B seemed to have lowered the oxidation resistance—as judged from the time to failure of the above probes when held at a fixed temperature. Probes sheathed in Nicrobell-A (Nicrosil-plus) survived 7 times longer than those in Nicrobell-B at 1200°C and twice as long at 1100°C.

Rusby *et al* [91] compared the high-temperature performance of type N probes having 3 mm diameter Nicrobell-A sheaths, and presumably satisfying the ID-MIMS criteria, with that of type K thermocouples in a conventional MIMS design using Inconel sheaths. The tests consisted of progressively increasing the temperature from 1000 to 1300°C or of cycling the probes

2.6. TYPES K AND N

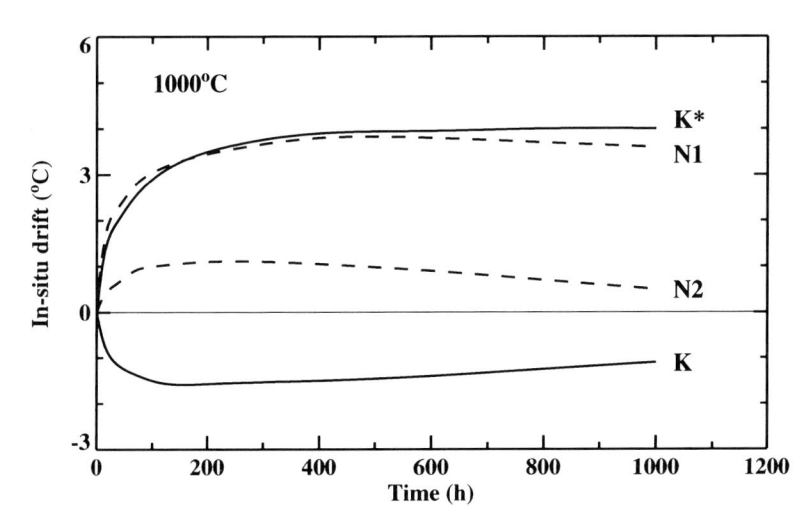

Figure 2.13: The *in-situ* drift in calibration as a function of time of 3.0 mm diameter ID-MIMS probes with Nicrosil-plus sheaths held at 1000°C, from ref. [84]. The thermoelements are type K (K, K*) and type N (N1, N2).

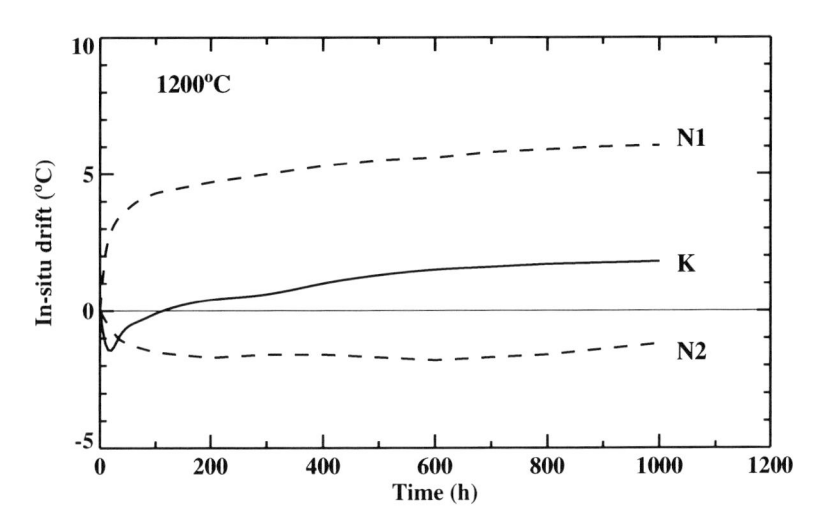

Figure 2.14: The *in-situ* drift in calibration as a function of time of 6 mm diameter ID-MIMS probes with Nicrosil-plus sheaths held at 1200°C, from ref. [84]. The thermoelements are type K and type N (N1, N2).

between temperatures in this range. Thus, their data are not comparable with the *in-situ* drift data given above. Nevertheless, as expected, the type N ID-MIMS probes performed better than the Inconel-sheathed type K.

There are a number of other articles [81, 92, 93] containing 'drift' data for Nicrobell-sheathed MIMS probes. These data are not included here—in that article [81] in which an experimental procedure is described the data do not relate to *in-situ* drift (see comment (f) on page 68) and it seems that the methods [93] used in the others were similar.

In one study [93], type K thermoelements within Nicrobell-sheathed MIMS probes failed prematurely at 1200 and 1280°C, e.g., in only 800 h at 1200°C for a probe diameter of 6.4 mm. It is difficult to account for this unusual behaviour, apart from suggesting the common problem of poor tip welds (section 2.6.3), since all probes failed by open-circuiting near the tip (see item (d) in the next section).

2.6.7 Availability of ID-MIMS probes

Quality assurance for the welds and the hermetic seals of MIMS probes (Section 2.6.3) is not specifically given by any probe manufacturer/fabricator, to my knowledge. Hence the only choice, for a user wishing to obtain an ID-MIMS probe, is to select a sheath that ought to satisfy the ID-MIMS criteria (page 56) and demand of the fabricator that adequate attention be given to the welds and seals. If the type K specification is required the Alumel needs to be a Mn-bearing alloy, and thus not the K* alloy mentioned above.

As to the sheathing alloy, there are three choices.

- Nicrosil-plus, but no longer available from MM Cables Pyrotenax. The same alloy is available as Nicrobell-A, the trade-mark of Nicrobell Pty Ltd.

- Other Nicrobell alloys. In 1992, Burley indicated [93] that Incotherm Ltd was developing four other alloys based on Nicrosil, for example, Nicrobell-B containing Nb for enhanced ductility.

- The alloy used by Hoskins Manufacturing Co in their 2300^{TM} MIMS cable [94]. The sheath contains two layers formed from strip and the probes are for continuous use at temperatures up to about 1260°C (2300°F).

With the advent of the ID-MIMS design, the approach to the use of thermocouples in difficult environments must also change. If the sheath of an ID-MIMS probe is unsuitable for a particular environment it is better to

2.6. TYPES K AND N

enclose it in a separate protective sheath than compromise its life and stability by changing the sheathing alloy, as is the practice for conventional MIMS thermocouples.

2.6.8 MIMS: type K or type N?

In deciding which of the two thermocouple types, K or N, performs better in the MIMS format, data obtained for bare-wire thermocouples should not be considered. The behaviour of thermoelements when sealed within a MIMS cable differs from that of the same alloys when 'bare', for obvious reasons—the causes of instability in Seebeck coefficient are different. Some authors have extrapolated from bare-wire data to the MIMS system, but this step is not valid.

Equally, it is unwise to form comparative conclusions about the ID-MIMS system from data obtained for MIMS thermocouples sheathed in conventional alloys.

As indicated in sections 2.6.3 and 2.6.4 the type K thermocouples, as a group, behave just as well as the type N when in conventional MIMS probes. This outcome is at least partly due to the dominance of other factors, such as, different sheathing alloys, with their differing levels of Mn (section 2.6.5), and the problems associated with welding and sealing procedures. Indeed, much of the poor stability and reproducibility of MIMS probes can be attributed to these causes.

On the other hand, these effects are overcome if the ID-MIMS criteria are satisfied (page 56), and the thermoelectric performance of the ID-MIMS probe is then due entirely to the behaviour of its protected thermoelements. Even then, the position is the same as that stated above—that type K and type N probes are equally stable, overall. Nevertheless, in the ID-MIMS system, there are some significant behavioural differences between the different thermoelements. What governs the behaviour of these probes is not so much whether the thermoelements are type K or type N, *per se*, but which of the following features applies.

For type K thermoelements the significant characteristic seems to be the composition of the negative leg, Alumel, and in particular the levels of Mn and Al. With roughly 1.5% of each of these elements (the K1 alloys, above) drifts of $< 2°C$ will occur at high temperatures (6 mm OD at 1000 to 1200°C) and, if the composition is optimised [86], ID-MIMS type K probes could be more stable. Conversely, if Mn and Al are absent, drifts in excess of 5°C will occur in 1000 h (section 2.6.6).

For type N thermoelements the level of hysteresis is the dominant factor. Hysteresis, occurring along the probe, especially in the range 500 to 1000°C, governs high-temperature performance, and the degree of hysteresis present is batch/source dependent. So, drift for the initial 1000 h can be as little as 1°C or more than 5°C, depending on this property, i.e., on the source of the alloys.

Hence, in summary

(a) With conventional MIMS sheathing (section 2.6.4), or if the integrity of the weld and seal cannot be assured, instabilities of 10 to 30°C are likely to occur in 1000 h (6 weeks) at temperatures from 1000 to 1200°C and the behaviour will be neither predictable nor reproducible. Any differences that may result from choice of thermocouple type, K or N, will be lost; and the probes are likely to fail prematurely at these temperatures.

(b) For optimum performance choose probes that satisfy the ID-MIMS criteria (page 56) and contain either Mn/Al-bearing type K thermoelements or type N that have been tested and shown to be low in hysteresis.

(c) In the ID-MIMS configuration, given the differences in the Mn/Al levels in the type K and hysteresis in the type N, the two types behave equally well, overall.

(d) If the physical integrity of an ID-MIMS sheath fails (poor weld, becomes porous, etc) it no longer satisfies the ID-MIMS criteria, its thermoelectric behaviour is likely to be poor and it will fail rapidly at high temperatures. In this faulty condition, type N thermoelements would survive longer because of their improved oxidation resistance.

(e) Ultimately, the choice of thermocouple type will be influenced by the existing instrumentation and associated extension leads. In a properly fabricated MIMS probe the upper temperature limit for the two types is the same—it is dictated mainly by the sheath.

(f) Published 'drift' data may not necessarily refer to *in-situ* drift. In some cases, for example in references [81] and [95], the data have been obtained by removing thermocouples at regular intervals from test/field furnaces for calibration at shorter immersion in a laboratory furnace. The observed drift in calibration is then a measure of the change in Seebeck coefficient that has occurred in the hot zone of the test furnace. Such values are of use to the experimenter, but neither the sign nor the magnitude of the change can be taken as an indication of *in-situ* drift— a point discussed in section 1.3 and demonstrated by the example of section 1.4.2. See also, Figure 2.7 and the discussion beginning on page 62.

(g) Be wary of the conclusions from any comparative study unless the various factors outlined above have been considered and controlled. Probes from different suppliers will have differing levels of Mn/Al, hysteresis, quality of insulation and quality of weld, etc. For example, quite unwittingly, high-hysteresis type N thermoelements could be compared with type K having optimum levels of Mn and Al.

(h) Even with well designed comparative studies the data apply only to the specific test conditions used, because of the dependence of Seebeck coefficient on time, temperature and diameter (section 2.6.6). For a particular set of circumstances—test temperature, atmosphere, probe diameter, etc—one thermocouple type could be judged significantly better than the other, yet under different conditions the reverse may apply.

(i) More data are needed on MIMS thermocouples, especially from field trials conducted in different atmospheres. The data should be interpreted in the light of the above factors.

2.7 Thermocouples based on Constantan (types E, J and T)

Thermocouples of type E, J and T are limited in their range of applications, when compared to those of type K or N. For example, and as a guide only, the use of type E thermocouples should be restricted to temperatures from −250 to 850°C, the type J from 0 to 760°C, and the type T from −180 to 400°C. Nevertheless, they do have special characteristics that are useful in some situations. The type E thermocouple has a particularly high Seebeck coefficient, the type J is usable in reducing atmospheres and the type T has Cu as a thermoelement. The advantages of these characteristics are described below.

Thermocouples of type E, J and T have relatively large net Seebeck coefficients (Table 2.2), since their negative legs are of the alloy Constantan, which has the most negative value of Seebeck coefficient (Figure 2.2). Of the three, the type E thermocouple has the largest coefficient because Chromel, the alloy of its positive leg, has the highest Seebeck coefficient of all the common thermocouple alloys (Figure 2.1). The large coefficient of the type E thermocouple is its most useful characteristic. It is of particular advantage in the cryogenic region (section 2.8) and in electrically noisy applications where a high signal-to-noise ratio is desirable. Because the positive leg is Chromel, hysteresis in the type E thermocouple is of a similar magnitude to that in the type K. Hysteresis also occurs in the type J and T thermocouples, but is somewhat less.

The positive leg of the type J thermocouple is commercially pure Fe, and its thermoelectric properties, being sensitive to the $\sim 0.5\%$ impurity level, are somewhat variable. As a consequence, the Constantan alloy is not thermoelectrically interchangeable with the Constantan used for types E and T, since its composition is adjusted to match each source of Fe. Fe undergoes a magnetic transformation at 760°C and an $\alpha - \gamma$ crystal transformation near 910°C, which causes a discontinuity in Seebeck coefficient (Figure 2.1). Furthermore, Fe rusts in moist atmospheres and is not recommended for use below 0°C because of potential embrittlement [12]. The type J thermocouple can be used in reducing conditions to 760°C with a similar stability to that in air.

Copper has a small value of Seebeck coefficient at room temperature (Figure 2.1), and little inhomogeneity. As a result, the type T thermocouple is particularly useful in the differential mode (section 3.9). The thermal conductivity of Cu is high (Table 2.4) and, as a result, the sum of the conductivities for the type T thermoelements is roughly 10 times that of the double-alloy thermocouples, the types E, K and N. This is a problem in those applications where heat conduction to or from the tip is likely to affect the temperature of the object being measured. Oxidation of the Cu leg sets an upper limit of about 350°C to the use of type T thermocouples in air, although the effect of oxidation on emf is small [12]. In a vacuum or in an inert or reducing atmosphere the type T thermocouple may be used up to 400°C, the upper limit of the reference tables.

There is little data available on the stability of these thermocouples. *In situ* drift data [47] for 1000 h suggest that type J thermocouples are as stable as type K up to about 600°C and the type E is as stable to about 800°C. At 870°C, however, a study [47] of 3.3 mm bare-wire thermocouples found drifts of −12 and −3.5°C in 200 h for type J and E thermocouples, respectively, and 1.3°C in 1000 h for type K. It was also noted in this study that after 1000 h at 760°C a 3.3 mm type J thermocouple was completely oxidised and about to fail. Another study [96], of 1.6 mm bare-wire type J and E thermocouples at 777°C, found drifts of −23 and −2.5°C, respectively, in 600 h. By comparison, a type K thermocouple of the same diameter drifted 1°C.

2.8 Thermocouples for very low and very high temperatures

It is generally not possible for alloy manufacturers to select or adjust thermocouple alloys to comply with the standard emf tolerances both above and below 0°C in the same thermocouple. Hence, wires obtained for use above 0°C are unlikely to comply with the sub-zero tolerances, and vice versa. Of

2.8. OTHER THERMOCOUPLES

course, the compliance or otherwise to the emf tolerances is of concern only if small calibration corrections are required—'out-of-spec' thermocouples will function just as well.

The Seebeck coefficient of most thermocouples decreases with temperature below 0°C (table on page 27) and eventually becomes impracticably small. For example, the coefficients of type T, K and N reduce to $10\,\mu\text{V K}^{-1}$ at 39, 47 and 74 K, respectively, and this effectively sets the lower limit of usefulness. Of the conventional thermocouples, type E is the most useful, as it has the highest Seebeck coefficient. Its coefficient is roughly twice that of the types T, K and N around 50 K, and it falls to $10\,\mu\text{V K}^{-1}$ at 24 K.

In the cryogenic region, below about 90 K, the Chromel versus Au 0.07at.%Fe, the 'gold-iron' thermocouple, is particularly useful. The addition of small quantities of Fe to Au produces a negative peak in Seebeck coefficient and thus, when matched to Chromel, the thermocouple has a high and constant Seebeck coefficient over a large temperature range. It is $18\,\mu\text{V K}^{-1}$ to ±10% from 10 to 150 K, and falls to $10\,\mu\text{V K}^{-1}$ at 2 K. Reference tables for this thermocouple appear in reference [11] and the absolute Seebeck coefficients for those thermocouple metals useful below 0°C are plotted in Figure 2.15.

Thermal conductivity values for 20°C, given in Table 2.4, show that pure metals, especially Cu, tend to conduct more heat than do the alloys. As a consequence, the type T thermocouple conducts 10 times as much heat as the double-alloy, type E thermocouple. In the cryogenic region the difference is even greater; below 30 K the type T thermocouple conducts 200 times as much as the type E, of the same diameter, and Chromel versus AuFe conducts 14 times as much [97]. In many instances, thermal conduction poses a serious problem in measurement and if Cu or AuFe wires are used small diameters and special anchoring/wrapping techniques are required. For further information on cryogenic thermocouples see references [97, 98].

The Pt-based thermocouples allow measurements of temperature to about 1750°C, and above this temperature the most useful thermocouples are those based on W and Re. There are several thermocouple combinations in this category and a general summary of their properties and development is given in reference [5], pages 176–197. The metals in this group are W and various alloys from W 3Re to W 26Re and, to a varying extent, the main difficulty with them is that on heating above their recrystallisation temperatures (either in use or in an effort to improve their emf stability) they become brittle.

The recrystallisation temperature of W is 1200°C and it has the greatest embrittlement problem, whereas W 26Re contains enough Re for there to be no significant problem of this kind. When the intermediate alloys, such as W 5Re, are stabilised by heat treatment, they become brittle and may

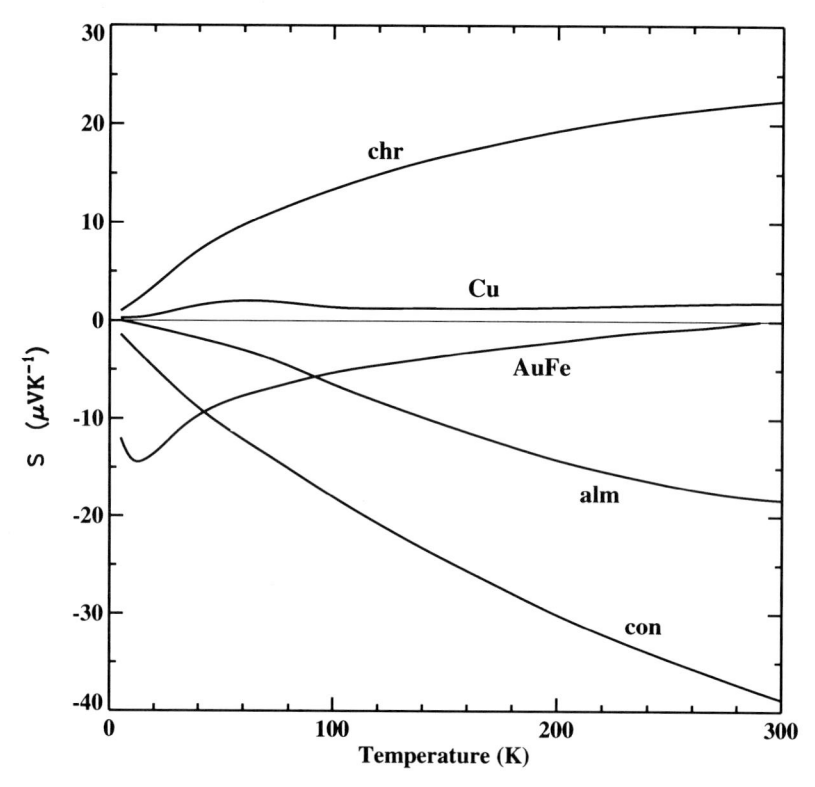

Figure 2.15: Seebeck coefficients, S, over the range 5 to 300 K for the metals Chromel (chr), Alumel (alm), Constantan (con), Au 0.07at.%Fe (AuFe) and Cu. Data were calculated using Seebeck coefficients relative to Pt [12], the coefficient for Chromel versus AuFe [11] and the absolute coefficient of Pt (page 214).

break on handling, though not as easily as W. This is overcome to a large extent by the addition of various dopants to the W and Re starting powders. The dopants are almost entirely eliminated during processing to produce a microstructure with enhanced ductility. The thermocouple combination W 5Re versus W 26Re is possibly the best choice and both wires are normally supplied in a thermoelectrically stabilised form. The combination is sometimes referred to as the **type C** thermocouple.

The W 5Re versus W 26Re thermocouple should not be used in air. In a vacuum, an inert atmosphere or in dry hydrogen its upper limit for reliable operation is considered to be 2760°C [11], which is well below the melting point of either thermoelement (Table 2.4). Reference tables are available [99] and its Seebeck coefficient is reasonable—at 0, 1000 and 2000°C it is 13.5, 18.4 and $12.2\,\mu\text{V}\,\text{K}^{-1}$, respectively. The recommended manufacturing tolerances are ±4.4°C or ±1%, whichever is the greater, up to 2315°C [99].

Much of the early developmental work was done at Hoskins Manufacturing Co [100]. This involved the development of suitable techniques in powder metallurgy, considerations of room temperature ductility and the development of alloys as compensating leads. A compensating lead must have a similar Seebeck coefficient to that of the thermocouple it replaces, in the region 0 to 200°C or so—see section 3.4.

Some long-term drift data are given in reference [41] for 0.5 mm diameter wires of W 5Re versus W 26Re in high-quality alumina, twin-bore insulation. At 1330°C the drift, −22°C in 10 000 h, is the same for wires in argon and in vacuum, the changes being similar to those found for thermocouple types R and B under the same conditions (see table on page 40).

2.9 Tip formation for bare-wire thermocouples

For a thermocouple to function correctly the wires must form part of a circuit that has electrical continuity, e.g., they must be in electrical contact at the tip, or tips if a differential thermocouple. Further, the tip may need to survive vibration, oxidation and handling. It may also have constraints on its size and mass. To achieve these objectives it is usual to weld the wires together or join them by the melting of a third metal, such as a solder.

For the rare-metal types (B, R and S) the tip should be formed by simply crossing one wire over the other, with an overhang of maybe 0.5 to 1 mm, and welding without flux in a small oxidising flame. As a fuel gas, 'town gas', 'natural gas', propane, butane or hydrogen may be used. The gas is mixed with oxygen in a small glass-blower's torch, having a jet diameter of 0.6 to 0.8 mm. The flame is best directed onto the alloy with the highest melting point and, on melting, the metals flow back to form a small bead. Goggles are required to reduce the light intensity.

Base-metal thermocouples, used up to 600°C, may be joined with silver (hard) solder. This is facilitated by making an open spiral, 15 mm in diameter and 50 to 70 mm long, wound from 1 to 1.5 mm Nichrome[1] wire. Beyond one end of the spiral the wire should have a 20 to 30 mm straight section and a closed loop at its end, of 3 to 5 mm internal diameter. By holding the spiral at the other end, the loop may be heated in a Bunsen flame and silver solder, in wire form, fed into the loop with a suitable flux. In use, the loop with its solder is heated till glowing, and held in the edge of the flame to maintain the solder in a molten state. The thermocouple tip, twisted and coated in flux

---
[1]reg. trademark of Driver Harris.

(paste), is then immersed in the liquid solder and withdrawn. Finally, the tip is cleaned of flux and cut to the desired size.

On forming a twisted thermocouple tip, it is important to remember that the thermocouple 'measures' the temperature of its first point of electrical contact, as seen from the CJ end. When a thermocouple is new and unoxidised, this point will be where the twisted wires first touch. Later, this effective tip may move as the wires expand and oxidise. In either case, it may not be the point assumed by the operator as the point of measurement. If this is a problem it may be resolved by having the wires touch first at the solder or weld and by having a short tip length.

For heavy-gauge base-metal wires an oxy-acetylene torch is usually used to weld the tip. The wires are held in a mechanical vice and given a tight twist of 2 to 3 turns. The thermocouple tip is then heated and placed in powdered borax or boric acid to give it a coating of the flux. It is then heated, with the aid of protective goggles, in a slightly reducing flame to form the weld. When cooled to below 'red heat' it may be dunked in water to loosen the flux. Striking with a hammer will then remove further flux and reveal a poor weld, if one exists. To finish, the weld may be cleaned with a wire brush and hot water.

2.10 Insulation and protection for bare-wire thermocouples

Individual thermocouple wires should be separated from each other by a highly insulating material, yet not be contaminated by it. An outer sheath or tube may also be required to protect the thermocouple from the environment. It should be remembered that sheathing may disadvantage temperature measurement because it tends to isolate the thermocouple from the site of interest. Sheathing also increases thermal conduction to ambient, it introduces a potentially stagnant air space and greatly increases the thermal response time.

Sheathing of Pt-based thermocouples poses another problem. These thermocouples are very susceptible to contamination and each must have one continuous length of alumina twin-bore insulation along its heated length. In other than clean air it is usually further protected within a closed-end alumina sheath and a second, metallic sheath may also be required. Problems arise because of the inflexibility of the ceramic components. They break on jarring or when transported, because they are unrestrained within the outer sheath, and also break if the probe warps on heating.

2.10. INSULATION AND PROTECTION

2.10.1 Flexible insulation

Thermocouples in flexible insulation are usually colour coded as a means of identifying both their type and their electrical polarity. This is fine for someone handling new material from a known source, but for thermocouples in service, particularly if the country of origin is not known, the code may be misleading. Red signifies the positive lead in Germany and the negative lead in USA, whereas blue denotes the negative in UK. Colour on the other wire is intended to identify the thermocouple type.

For laboratory use, particularly on small diameter wire, **enamel** is useful up to 100°C, but it has poor abrasion resistance. Flexible and economic insulation is afforded by **PVC**, which is resistant to abrasion, water, oil and many chemicals. It is dissolved by some mineral solvents, such as benzene, swells in acetone and is decomposed by concentrated oxidising acids. All forms of PVC are degraded by heat and light and as such have poor service if exposed out of doors. Decomposition in air begins at 100°C and the material is flexible down to −30°C.

When excellent abrasion resistance is required **nylon** is often chosen. It is affected little by hydrocarbons, oils and many other organic liquids, concentrated alkalis and dilute acids. It is decomposed by concentrated or hot dilute mineral acids, by oxidising agents and halogens. It experiences slight water absorption and flexibility is retained down to −60°C. Nylon is serviceable to 150°C and most forms degrade rapidly by oxidation above 200°C. In the absence of oxygen decomposition begins near 300°C. For corrosion resistance **Teflon**[2] is chosen. It is decomposed by molten alkali metals and prolonged use in fluorine, and it is soluble in no known liquid at room temperature. It is still flexible at −80°C and is not completely brittle at −180°C. Teflon FEP (fluorinated ethylene propylene) may be used continuously to 200°C and Teflon PTFE (polytetrafluoroethylene) to 260°C. Both Teflon's may be taken ∼50°C beyond these limits for a single reading. For higher temperatures **Kapton** (polyimide) may be suitable, for example, Kapton-H[3] may be used continuously to 300°C, and for ∼100 h at 350°C. It can be used up to ∼400°C for a single measurement.

There are several fibrous materials available, such as **fibreglass**, with good survival at 500°C, **silica** braid for use up to 1100°C, and woven **ceramic-fibre** insulation, usable to the upper temperature limit for type K and N thermocouples. These fibrous materials tend to have inherently poor abrasion and moisture resistance, although fibreglass is somewhat better than the others. Often a coating or impregnation is added that gives fair resistance to abrasion and moisture up to maybe 150°C. If a thermocouple with this

---
[2] reg. trademark of Du Pont USA.
[3] reg. trademark of Du Pont USA.

form of insulation is to be used above $\sim 300°C$ problems may arise unless these chemicals are removed by an isothermal anneal at 300 to 400°C (see section 5.7).

2.10.2 Rigid insulation

Lengths of single or twin-bore insulators in various grades of **porcelain** (up to maybe 1400°C) are available as well as high purity recrystallised **alumina** (up to 1800°C), the latter being required for platinum-based thermocouples. **Magnesia** is limited to about 2000°C because of its low electrical resistivity, and for higher temperatures **beryllia** is satisfactory. For temperatures to 1000°C **silica** may be used and for the lower temperature region (up to 500°C) various forms of **glass** have their use. The thermal shock resistance of silica is excellent and that of the others is usually adequate.

The ceramic beads used with base-metal thermocouples should not be tight. Most thermocouples need free access to air, to avoid stagnant, and thus partially-oxidising conditions. For Pt-based thermocouple wires the ceramic insulation should be one continuous length over all sections of the thermocouple likely to be at temperatures above 600°C (section 2.5). Ideally, for laboratory use, the alumina insulation should be a single length to below 60°C, where insulation may be continued in PVC or Teflon.

2.10.3 Protection tubes and sheaths

Metallic tubes

There are a variety of metallic sheaths available with various upper temperature limits (given below in brackets). The limits set by oxidation are 20 to 50°C above or below these figures, depending on whether the application involves intermittent or continuous use, and on the mechanical stability of the oxide scale. Often, the limit for intermittent use is below that for a continuous one, an exception being 446 stainless steel, which develops an adherent scale at high temperatures. The porosity of a metal tube may become a problem above $\sim 800°C$ and, then, thermocouples require the further protection of an intermediate ceramic sheath.

Carbon steel (550°C) is an economic choice for 'moderate' temperature applications provided the atmosphere is non-corrosive. It is used with molten non-ferrous metals, galvanising, tinning and many petroleum applications. **Wrought iron** (650°C) also requires a non-corrosive atmosphere and is used in such processes as annealing, drawing and tempering etc. **Cast iron** (700°C) is used in the chemical industry (withstands sulphuric acid and caustic solutions)

2.10. INSULATION AND PROTECTION

and with daily applications of white wash is used for molten aluminium or die-cast metals.

304 stainless steel (900°C) has the nominal composition 70Fe 19Cr 10Ni and has good resistance to a wide variety of corrosive media. It has general use in wet-process applications, such as steam lines, oil refineries and chemical solutions. It is recommended for use in nitric acid, does tolerably well in sulphuric, phosphoric and acetic acids but does poorly in the halogen acids. The corrosion resistance is improved further with additional molybdenum in the form 68Fe 17Cr 12Ni 2Mo, the nominal composition of **316 stainless steel** (900°C). This alloy is resistant also to the complex sulphur compounds used in pulp and paper processing and resists pitting by phosphoric and acetic acids.

310 stainless steel (1100°C) has the approximate composition 52Fe 25Cr 20Ni. At high temperatures, it has good resistance to oxidising and carburising atmospheres and is widely used where sulphur dioxide gas is encountered. It resists fuming nitric acid at room temperature and fused nitrates to ~ 420°C. **Incoloy** (1100°C) has a similar composition, being 46Fe 20Cr 33Ni, and is used for high temperature heat treatment applications as well. **Nickel** (1100°C) has a similar temperature limit but its principal use is in hot caustic and molten-salt baths, e.g., in molten potassium cyanide. It should not be used in the presence of sulphur above ~ 550°C. If sulphur corrosion is a problem **446 stainless steel** (1100°C) may be the solution. Its composition is nominally 74Fe 25Cr and it has excellent corrosion resistance, but its mechanical strength is low when compared with, say, Inconel. Its uses include thermocouple protection in high-temperature hardening and nitriding salt baths, vitreous enamelling, non-ferrous metals smelting and blast furnaces.

The Inconel's have good high-temperature properties and, unlike the above alloys, contain more nickel than iron. **Inconel**[4] **600** (1150°C) is nominally 73Ni 16Cr 10Fe and **Inconel 601** (maybe 1200°C) is 61Ni 22Cr 13Fe 1.5Al. The latter alloy forms the more stable oxide scale because of the additional chromium and aluminium, yet both versions do well in high-temperature applications. These include carburising, nitriding and salt baths as well as blast furnaces, gas generators and ceramic kilns. Inconel should not be used in the presence of sulphur above ~ 550°C.

It is my understanding that the **Nicrosil**-based alloys, Nicrobell-B and Nicrobell-C (see Section 2.6.7), are available as protection sheaths for use up to 1300°C.

The Hoskins' alloys Chromel A (78.4Ni 20Cr 1.0Si 0.5Fe) and Chromel AA (68Ni 20Cr 2.0Si 8.3Fe) are designed specifically for use as heater wires and tapes at high temperatures in common atmospheres. For example, the latter alloy works well in both carburising and sulphurous conditions. Unfortunately,

---
[4] reg. trademark of the INCO group of companies.

they are not as yet available as sheaths or in tube form, a situation that also applies to the equivalent alloys from other manufacturers.

Ceramic tubes

Most of the ceramic materials used for the sheathing and/or protection of thermocouples are oxides (Table 2.9). For conductivity and expansion data on some of these materials see Table 2.4 on page 29.

Table 2.9: Composition of ceramics used for sheathing and their approximate upper temperatures of use (very dependent on the application, e.g., whether vertical or horizontal or mechanically loaded). All materials listed, except SiC (porous), are impervious to gases.

Ceramic	Composition	Colour	Upper Limit (°C)
Alumina	Al_2O_3	white	1800
Aluminous porcelain	$SiO_2 + Al_2O_3$	white	1500
Beryllia	BeO	white	2200
Magnesia	MgO	white	2300
Mullite	$\sim 3Al_2O_3.2SiO_2$	buff/brown	1600
Silica	SiO_2	clear	1300
Silicon carbide	SiC	grey	1600

The thermal shock resistance of silicon carbide and silica is excellent, that of mullite is good, alumina is fair and magnesia, poor. On the other hand, the mechanical strength of alumina is high, greater than that of either mullite or silicon carbide.

Silicon carbide has excellent abrasion and wear resistance, is highly resistant to the cutting action of flames and gases and to the corrosive effects of sulphur dioxide. However, it is porous. Thus, if the surrounding gases could affect a thermocouple within the carbide sheath a further, impervious sheath (e.g., of alumnia) is required. Thermocouple protection sheaths, such as those of Carborundum[5], are usually of SiN-bonded silicon carbide. As a consequence, its thermal conductivity is about 10% higher than that given for SiC on page 29.

Being available in pure form (> 99.7%), **alumina** is recommended where clean working conditions are essential, especially for the insulation and protection of Pt-based thermocouples. As well as having high mechanical strength and being impervious to gases, it is resistant to slags, chemical attack

---
[5] trademark of Saint Gobain Co, France

2.10. INSULATION AND PROTECTION

and reducing gases. Alumina is used for induction melting and other high temperature processes. For even higher temperatures **beryllia** can be used. It has excellent resistance to thermal shock but is expensive, its fumes and powder are highly toxic and it may react with other oxides.

Mullite is an alumino-silicate named after the Isle of Mull where it was first mined. Even though its composition is usually given as that in Table 2.9, it varies, depending on the source and its manufacture, from having 62% to about 75% Al_2O_3.

Another high-strength ceramic is **Sialon**, which I understand to be a silicon-nitride-based material with interspersed atoms of Al. It is expensive, but has a very high thermal shock resistance and is resistant to erosion by molten aluminium.

There is also a range of **cermet** tubes available. 'Cermet' is a generic term for materials having two interlacing continuous phases, one of metal and the other of ceramic. It is a high-strength, corrosion resistant material with good thermal shock resistance and is produced by combining 60 to 80% Mo with Al_2O_3 or $\sim 75\%$ Cr with Al_2O_3 and SiO_2, for example [45]. Cermets are useful in molten metals, such as steel (Mo cermet) and Cu or its alloys (Cr cermet).

Chapter 3

Thermocouples in Use

3.1 Introduction

As discussed in section 1.3, every conductor generates emf wherever it conducts heat. Both of these effects, the generation of Seebeck emf and the conduction of heat, are responses to the local temperature gradient. Thus, all the metals in an electric circuit, including extension and connecting wires, switches, terminals and circuit boards, will produce Seebeck emf. The total emf in the circuit depends on as many temperatures and Seebeck coefficients as there are metals. Therefore, if a meaningful measure of temperature is to be obtained at any one point the emf contribution from extraneous materials should be limited.

In its simplest form, a thermocouple circuit will produce Seebeck emf because the temperatures at two sites differ, and to use the signal as a measure of one temperature we need knowledge of the other—that of the reference-junction or cold-junction. Indeed, it is useful to realise that, in one sense,

- a thermocouple transfers the problem of temperature measurement from one site to another—from a site that is relatively difficult to measure to one that is easy.

We need to be clear as to the site of the unknown but desired temperature (tip) and the position of the reference or cold junction (CJ). Between the two, all components of the circuit must be so chosen that they generate emf according to the same emf-temperature relationship. In addition, that part of the circuit beyond this important tip-to-CJ zone must not produce significant Seebeck emf.

Ideally then, we need a single continuous length of homogeneous thermocouple cable from tip to CJ—a thermocouple free of discontinuities and foreign materials. Also, we need to know the temperature of the CJ and to avoid any emf between it and the signal-measuring/processing circuit. In practice, a thermocouple circuit often contains more components than is ideal—it may have switches, plugs and sockets, and other wires. It is thereby more interesting, as it has a greater variety of error sources, but more of this later, in section 5.7.

The next section of this chapter considers the effect of using a real thermocouple—one that is inhomogeneous (a description that is especially appropriate if at least part of the thermocouple had been in use above about 250°C). The section which follows deals with methods of coping with the cold-junction temperature and the remaining sections cover the effects of disrupting the thermocouple by introducing 'extension leads' and switches.

3.2 Real thermocouples

'Real' thermocouples are not thermoelectrically homogeneous, they have a significant level of inhomogeneity that may increase with use. The concept of inhomogeneity was introduced on page 13 as a growth in thermoelectric signature, and a practical consequence of such change was calculated in the example beginning on page 21.

Further insight can be gained from data of the type depicted in the figures on pages 37 and 46. Here, the changes that occur in Seebeck coefficient are given as functions of the temperature at which the changes occur. Effectively, each curve shows the local change in Seebeck coefficient developing along the thermocouple. A thermocouple with a tip temperature of 1100°C, say, will experience along its length the full range of temperatures from ambient to 1100°C. Somewhere along a type K thermocouple there will be a region at ~300 to 600°C, where the Seebeck coefficient will have increased during the time spent at these temperatures (see figure on page 46). Similarly, further towards the tip, there will be a region at 800 to 1100°C where the Seebeck coefficient will have increased or decreased, depending on the format (bare-wire or MIMS). It is because of these changes that the thermocouple signal drifts in time. The above-mentioned figures indicate the regions from which the contributions to drift occur as well as the relative significance of each source of change.

Because the various contributing processes occur at differing rates and their contributions differ in sign, the net drift in thermocouple signal often fluctuates in its direction of change. For example, the emf's of type K and N2 thermocouples, of Figure 2.9, first decrease with time and then increase.

Similar curves can be constructed for all types of thermocouple. For example, in types R and S the Seebeck coefficient undergoes reversible change

3.3. THE COLD JUNCTION 83

in two regions, one peaking at around 450°C, like that of type K, and the other at around 800°C (see Figure 2.5 on page 37). It also undergoes a significant irreversible change at temperatures above about 1300°C, if under clean laboratory conditions, and at lower temperatures if not. Fortunately, the changes in Pt-based thermocouples are almost two orders of magnitude smaller than those in the Ni-based types.

So, in general, the Seebeck coefficient at every point along the length of a 'real' thermocouple changes with time by an amount dependent on the local temperature, and the net effect is a drift in thermocouple signal. Equally, and just as important, the changes represent a continually increasing degree of inhomogeneity.

Notice that the degree of inhomogeneity is **not** reflected by the extent of drift witnessed at a fixed immersion depth because changes in Seebeck coefficient that are in opposite directions add to inhomogeneity but tend to cancel in their contribution to drift. The full effect of inhomogeneity will not be experienced until the temperature profile along the thermocouple is changed, for example, by a change in the depth of immersion.

A real thermocouple is relatively homogeneous when new—it begins service by producing a signal that depends on two temperatures only, those of its tip and CJ (to better than 99.9%—inhomogeneity being less than ±0.1%, see page 13). With use at temperatures above 250°C, the thermocouple signal will generally develop a growing dependence on its longitudinal temperature profile.

The growth in inhomogeneity, and the subsequent dependence of emf on depth of immersion, can sometimes be minimised by an appropriate anneal. This is especially so if the intended use involves temperatures up to, but not significantly higher than, the temperature at the peak for reversible change, e.g., 450°C in type K thermocouples (see page 46).

3.3 The cold junction

The pair of connections/junctions where the generation of thermocouple emf nominally finishes is collectively referred to in the singular, as the reference or cold junction (CJ), because their temperatures should be the same (see section 1.3). In practice, the need to measure the CJ temperature is handled in one of three ways:

1. by arranging the CJ to be at a known temperature, such as in a temperature controlled bath or in a fixed point. The simplest and most reliable fixed point is the 'ice pot', which establishes a region at 0°C and doesn't require measurement,

2. by separately measuring the CJ temperature with a suitable sensor, such as a thermistor or resistance thermometer element (RTD), or

3. by using a measuring instrument (section 3.5) that features automatic CJ compensation (ACJC), which in effect uses method 2 automatically and displays the compensated thermocouple signal in temperature units.

The first two methods are not compatible with an instrument using an active ACJC circuit. These methods usually require that thermocouple signals be measured in emf units. The first method is used in situations that require maximum accuracy and the third is the most common.

In each case, temperature is derived using the following procedure:

$$V_{T \to CJ} + V_{CJ \to 0} = V_{T \to 0} \quad (3.1)$$
$$\to T.$$

The first term on the left is the emf generated by the thermocouple, from tip (temperature T) to CJ, and the second is that equivalent to the CJ temperature, 'measured' by one of the above methods. The total emf is converted to temperature using reference tables or the equivalent relationship 'built-in' to an instrument.

An ice pot is a suitably contained mixture of crushed ice and water (preferably both distilled). It is a fixed point, and as long as water and ice are together in equilibrium, while exposed to an atmosphere of air, the temperature is 0°C. The container is usually a Dewar flask or an insulated stainless steel can and the thermocouple wires are connected to copper wires within clean, thin-walled glass tubes that are immersed in the container. A convenient system is to have two tubes, each containing mercury to a depth of about 20 mm. One thermocouple wire and its corresponding copper wire, both insulated, except within 5 to 10 mm of their ends, are then immersed in the mercury of each tube to form the CJ. An immersion into the ice/water mixture of 150 mm would guarantee a CJ temperature within ±0.01°C of 0°C for 0.5 mm Pt-based thermocouple wires connected to copper wires of a similar diameter, for example. An ice pot offers an easy way of achieving a uniform-temperature zone and hence of assuring that the CJ temperature is as assumed. It also has the advantage of having a known temperature, not requiring measurement.

There are 'automatic ice points' available. Their operation relies on the difference in density between ice and water to detect the ice point and on the Peltier effect (section 1.2) for controlling the temperature of a small work chamber. The chamber is uniform in temperature to maybe ±0.02°C and needs an occasional check on its temperature to ensure the device is operating correctly. Also, care must be taken in choosing wire diameters, to minimise heat conduction into the critical region: more care is needed than with a conventional ice pot because of the smaller depth of immersion.

If the CJ temperature is to be measured and corrected for, as in method 2 above, the temperature of the chosen sensor must equal those of the CJ ends. This is not necessarily easy. For example, many sensors are affected by 'self-heating', which increases their temperatures above that of their immediate surroundings.

3.4 Extension leads

Because of environment and temperature the thermocouple may need to be of heavy gauge or of an expensive material. It may also need to be insulated with special ceramics and a sheath may be necessary. Usually, it is impractical to extend a thermocouple from tip to CJ in this form because the CJ needs to be some distance away, at the instrument terminals where the ACJC monitoring occurs. It is usual to break the tip-to-CJ circuit into two parts, each with its own clearly defined role. That part adjacent to the tip, whose task it is to produce most of the signal, and the remainder of the circuit, used to extend thermoelectric generation to the CJ at the instrument terminals. The first part is loosely referred to as the 'thermocouple' and is often supplied with terminals at its open end, the **'head'**, to allow connection to the second part, the 'extension leads'. Obviously, extension leads should have thermoelectric properties that match those of the length of thermocouple it replaces. It would then produce an emf equal to what the thermocouple would have produced if instead it had extended over the same temperature interval.

There are two types of extension lead:

- Those known commercially as 'extension wires'. They have a composition nominally the same as the thermocouple and are usually covered in flexible insulation suitable for the temperatures they are designed for (typically up to 200°C). The manufacturing tolerances for base-metal types are about ±2°C. For Pt-based thermocouples extension wires are not available as such. Instead, for laboratory use, they may be 'extended' using thermocouple wire of the same type, e.g., wire salvaged from old, no longer useful, standards may be adequate.

- Those known commercially as 'alternate extension wires' or 'compensating wires'. They have compositions that differ from the thermocouple and are available for thermocouple types B, S and R, for the W/Re thermocouples (section 2.8) and for type K. Their manufacturing tolerances are large (see below).

In my view, compensating wires should not be used for base-metal thermocouples (section 2.2). For these thermocouples inexpensive extension wires are available, and any cost saving there may be in using compensating wires is outweighed by the unnecessary increase in error. For example, copper and Constantan are often supplied as compensating wires for type K thermocouples. They generate too much emf: the equivalent of 5°C at a head temperature of 100°C and 29°C at 200°C.

Compensating wires are available for the W/Re based thermocouples, with tolerances of ±0.11 mV for (head) temperatures up to 870°C [101]. This amounts to ±9°C when measuring a temperature near 2000°C, or ±0.45% for a W 5Re versus W 26Re thermocouple. For the type B thermocouple the combination Cu 1.5Mn versus Cu is used as compensating wires. Its emf, for head temperatures up to 200°C and a CJ temperature of 0°C, would match that of a type B thermocouple to within about $20\,\mu$V [102]. If the head temperature can be kept below about 50°C it is better to use a pair of Cu wires as the compensating lead. A pair of Cu wires produces no emf and that part of a type B thermocouple operating from 50°C, at the head, to 0°C (or 20°C), at its CJ, would produce no more than $3\,\mu$V (or $5\,\mu$V, respectively), equivalent to 0.3°C (or 0.5°C) for a tip temperature of 1000°C.

The compensating wires for type S and type R thermocouples are the same, viz, Cu versus Cu 3Ni 2Mn and referred to by the code SX. The manufacturing tolerances vary from country to country and are typically $\pm 60\,\mu$V for (head) temperatures up to 200°C, assuming 0°C for the CJ temperature. A study of many manufacturing 'heats' of SX alloys has been made in which their emf's were compared with the type S and type R reference tables [102]. For the temperature range 0 to 200°C the largest difference was $-60\,\mu$V. It occurred at 200°C and refers to the type R relationship. Further, it seems that if the CJ temperature of SX compensating wires is between 20 and 50°C, instead of 0°C, as is more likely in practice, the lead would contribute an error of $80\,\mu$V, for a head temperature of 200°C. This is equivalent to about 7°C for a thermocouple tip at 1000°C, for example.

It is useful to remember that the mismatch error introduced by compensating wires is usually reduced as the temperature interval, for which the lead is required to generate emf, is reduced. For example, if the head and instrument terminals are at the same temperature the lead will generate no emf, as required, and the associated error is zero. Examination of the data on SX alloys in reference [102] suggests that at 20°C the rate of change for the mismatch error is $\leq 0.6\,\mu$V K^{-1}. So, if the CJ and head temperatures differ by less than 10°C the error in using the lead is likely to be less than 6μV, equivalent to 0.5°C—assuming a Seebeck coefficient of about 12μV K^{-1}, e.g., a type R tip at 600 to 1000°C (see Table on page 27).

3.4. EXTENSION LEADS

Conversely, the error can be considerable. When compensating wires are used for a platinum-based thermocouple it is advisable for routine checks to be made on the head temperature, to limit the associated errors. A simple technique using the human hand is recommended. If the hand can be held on the thermocouple head the terminals are unlikely to have a temperature in excess of 60°C, and the error due to the lead will probably be less than 2°C for a CJ temperature of about 20°C.

The use of an ice pot at the CJ of compensating wires for type R or type S thermocouples is unwise. An ice pot is usually employed when an improvement in accuracy is sought, yet with a compensating cable it merely increases the need for the cable to generate emf and hence error.

If short type R or S thermocouples must be used for accurate laboratory work they should be extended with matching wires of Pt and PtRh, insulated in PVC or Teflon. The extension wires may be attached by welding, with suitable clips or pressed together with a screw in a suitable connector block. For example, the block could be of perspex with four drilled holes to take the wires and two tapped holes for nylon screws. For each leg, the two wires pressed together by a screw would preferably be at right angles to each other and to the screw.

Since the active zone, from tip to CJ, is usually in two sections joined at the head the word 'thermocouple' may be used ambiguously. There is the thermoelectric view, which would define the thermocouple as that producing the measured signal, and therefore includes all the circuit. Ideally, this should be a single continuous cable, and the fact that it is discontinuous at the head doesn't alter the approach. On the other hand, 'thermocouple' commonly refers to the business end of the circuit, an assembled probe with a head to which an extension lead is attached.

3.4.1 Extension lead polarity

Because of uncertain colour coding, on either the thermocouple or the extension lead, or both, the polarity of the wires may be in doubt. If the thermocouple is incorrectly connected, in relation to the instrument, it would normally be noticed on its first use. The instrument would show a decrease in temperature while the furnace was heating up, for example. On the other hand, an incorrectly connected extension lead is not so easily noticed: it would contribute an emf component of the wrong sign and often of relatively small magnitude. For a head temperature 20°C above that of the CJ a type K thermocouple would measure low by 40°C, and a type R by ~ 20°C, if the polarity of the extension lead was incorrect. The polarity of the lead may be checked by one of the following methods.

- Disconnect the extension lead from the head and form a temporary junction by clamping or twisting the wires together. If this junction is heated correct polarity is indicated when the instrument registers a temperature increase. If this method is impractical through lack of access, for example, the following alternative may be used.

- With the thermocouple and extension lead still connected at the head, a forced change in the CJ temperature, without affecting the temperature of the instrument terminals, will indicate the polarity of the lead. Simply disconnect the extension lead from the instrument and insert two fine-gauge copper wires so that the temperature of the newly formed CJ (extension wire/copper junctions) can easily be changed without affecting the ACJC network. Heat these junctions simultaneously, using the fingers or a small flame, and if the polarity is correct the instrument will register a fall in temperature. Alternatively, add some ice and look for a temperature increase.

3.4.2 Extension lead calibration

An extension lead may be given a single-point calibration—at 100°C, say—to check that it is within tolerance. Alternatively, the lead may be calibrated from ~ 0°C to the upper temperature likely to be encountered by the head in its intended application, the head being the set of connections where the thermocouple proper joins the extension lead.

If the temperature range to be experienced by the head and CJ in a particular measurement setup is restricted, e.g., a head temperature between 40 and 50°C and CJ, 20 to 30°C, the calibration data for the lead may be used to correct any such measurement. The correction to be applied is the reported correction for a temperature equal to the mean value expected for the head (45°C in this example) minus that for the CJ (25°C). Moreover, the uncertainty in the use of the correction can be gauged from the report by assuming possible variations in the head/CJ temperature combinations, e.g., from 40°C/30°C to 50°C/20°C.

If no restriction is placed on the head and CJ temperatures, other than limiting the head to a nominal maximum, e.g., 100°C, a correction for the lead cannot be applied. What could be done in this case is to estimate from the calibration report the range of errors possible for the nominal limit, assuming the CJ temperature lies between 0 and 40°C.

If the maximum likely error is considered excessive the extension lead could be replaced or the calibration data used to find a more suitable upper limit for the head temperature.

It is worth remembering that the use of an extension lead implies the assumption of compliance with the relevant reference tables—that its calibration error, $V - V_{ref}$, for all combinations of CJ and head temperatures, is zero. The assumption is tantamount to the acceptance of a systematic error (see section 5.3 on page 156). An estimate of the possible extent of such error, the corresponding component uncertainty (see note on page 172), may be taken as the manufacturer's tolerance or, for a smaller estimate specific to the lead being used, assessed by the calibration procedure discussed above.

3.5 Measurement of thermocouple emf

Three steps are involved in converting emf, developed by a thermocouple and its extension lead, if applicable, to temperature. The steps, with the most reliable and accurate methods of achieving them, are as follows:

1. Measure the emf developed by the thermocouple, best done with a digital voltmeter/multimeter (DVM) with $1\,\mu$V resolution, or better.

2. Establish or measure the CJ temperature, best done by placing the CJ at 0°C (in an ice pot or its equivalent, see page 84). Then, the CJ temperature doesn't require measurement. Otherwise, the emf equivalent to the CJ temperature needs to be added (arithmetically or electrically) to that of the thermocouple.

3. Convert the total emf to a value of temperature (hopefully close to that of the thermocouple tip!). The conversion is best done using software that incorporates the appropriate standard reference function.

Furthermore, the measurements may be approached as an exercise in emf measurement or by dealing directly in temperature units (see below).

Emf-measuring instruments (displaying in emf units)

A thermocouple generates an emf that is not dependent on its length or diameter, *per se*, and to avoid circuit losses the emf should be measured at zero current. Hence, measuring instruments suitable for thermocouple use have either a high-impedance input stage or, if requiring a significant current, as do moving coil types, the external circuit resistance should be fixed and specified. The stipulated external resistance includes that of the thermocouple and its extension lead and any change in thermocouple resistance, resulting from a change in its temperature profile, will affect the instrument response.

The most convenient instrument is the DVM, which is unaffected by thermocouple resistance (within limits). Moreover, it is relatively easy to obtain a DVM with a stability better than 0.01% and a resolution of $1\,\mu$V, or better, i.e., having an uncertainty significantly less than the smallest calibration uncertainty for a thermocouple (except for 'elemental' thermocouples: section 2.4).

As indicated above, once the emf developed in a thermocouple is measured, two further steps are required before the tip temperature is obtained. Firstly, the CJ temperature must be either established at a known value or measured and, secondly, the total emf must be converted to an equivalent temperature. The total emf—that developed by the thermocouple plus that equivalent to the CJ temperature (see page 19)—should be converted to temperature using the standard reference tables (or functions) given in Appendix B.

If switching between a number of different thermocouples is required the arrangement of Figure 3.5 is recommended.

Temperature-indicating instruments (displaying in temperature units)

Instruments designed for use with thermocouples and displaying in temperature units incorporate the three steps mentioned above, automatically. They perform these tasks with varying degrees of success. It is possible, in principle, for an instrument to perform the steps as accurately as can be achieved by the emf-measuring approach, above. However, it is likely to be more expensive and would be far less versatile.

Step 1 is achieved with an emf- or current-sensing circuit, preferably a DVM module, step 2 is performed with an ACJC network and step 3 via a representation of the standard reference functions. The resulting value of temperature is indicated by a display module that may be part of the same unit or distant from it.

In the latter case, the ACJC and emf-conversion networks may be incorporated in a separate unit, which is mounted directly on the thermocouple (no extension lead) and transmits a high-level signal (usually proportional to temperature) to a remote display unit. Because of the minimal distance between thermocouple and preamplifier, the system has a reduced sensitivity to electrical noise. On the other hand, there are size constraints, a greater and more variable temperature gradient around the ACJC detector and a potentially more hazardous environment. Thus, the uncertainty of measurement with such a system is somewhat larger than that obtainable with an integrated unit, which would be located in more stable and uniform ambient conditions.

If the emf-temperature relationship employed by a temperature-indicating instrument is functionally identical to the standard reference polynomial the conversion error is zero. This could apply to a microprocessor based instrument. It is more common, however, for an approximation to be used,

3.5. MEASUREMENT OF EMF 91

embodied in an analogue (linearisation) circuit, in the markings on an analogue scale or in a simplified version of the reference polynomials.

An ACJC system relies on a temperature sensor responding to the temperature of the input terminals of an instrument, whether these are within the confines of the instrument or remote from it. There is usually a difference in temperature between the sensor and the terminals, but this matters little so long as the difference remains constant (after warm up) and no input switching is involved (see page 93). It is then adjusted out during the ACJC check, which should be carried out either on a routine basis or just before a critical measurement.

For this adjustment, the instrument should be set to indicate 0°C when a thermocouple of the appropriate type is connected to the input and the tip adequately immersed into an ice pot. If the thermocouple is uncalibrated it will contribute an error of less than about 0.1°C if Pt-based, and about 0.5°C if it is a standard-grade base-metal type. For greater accuracy, premium-grade wire could be used or the thermocouple may be calibrated to a suitable accuracy and corrections applied (see page 151). As an alternative to using an ice pot, for this adjustment, the thermocouple tip could be thermally linked to a calibrated liquid-in-glass thermometer and the two placed at ambient. Then, the thermocouple would span a smaller temperature interval and its calibration would be less important. In this case the instrument should indicate the thermometer temperature.

The above comments relate to a stable temperature difference between the input terminals and the ACJC sensor. This condition may not apply if the temperature of the instrument is changing, such as when taken from one room to another. Then, because of differing time constants of the components, compensation for the CJ temperature will be in error by an amount that varies with time.

An instrument with ACJC is convenient in use, but it does have three avoidable error sources: those associated with (1) measuring the CJ terminals, (2) converting this temperature to emf and (3) converting the total thermocouple emf to a value of temperature. These processes are handled by the instrument and compromises are made. On the other hand, the first two error sources are avoidable with instruments that have an 'external ACJC' option—allowing the CJ to be placed at 0°C (the ACJC circuit is then bypassed)—and the third is avoidable if the emf-to-temperature conversion is done via the reference functions of Appendix B (by a microprocessor).

If switching between a number of different thermocouples is required and an instrument with ACJC is to be used, the CJ temperature for each of the thermocouples must be the same. Indeed, it must be that detected by the ACJC network. Consequently, the switching unit is normally integrated with the measuring instrument and arranged according to Figure 3.1. The combination, often referred to as a 'data logger', may allow the ACJC sensor to be remotely located. Then the arrangement of Figure 3.2 may be used. Alternatively, some instruments allow the option of an 'external ACJC', usually assumed to be 0°C, and the instrument connection is via a pair of Cu leads. In this case, the switching circuit of Figure 3.5 may be used.

3.6 Switching

If several thermocouples are to be measured by the one instrument a selector switch is used, often referred to as a 'scanner' and employed in 'data loggers'. It is commonly placed at their CJ's or inserted between the tip and CJ of each thermocouple, as shown in Figure 3.1. In the left-hand circuit of the figure the switch would be operated independently of the instrument, with the temperature of the single CJ being compensated for by the instrument. On the other hand, the switch on the right would form part of a measuring instrument or, if outside it, would normally be integrated with it in function—to allow the monitoring of the CJ and, thus, its compensation. In both cases, there exist small differences in temperature between the terminals of the switch, for which there are no corresponding Seebeck emf's. These differences contribute directly to the overall measurement error.

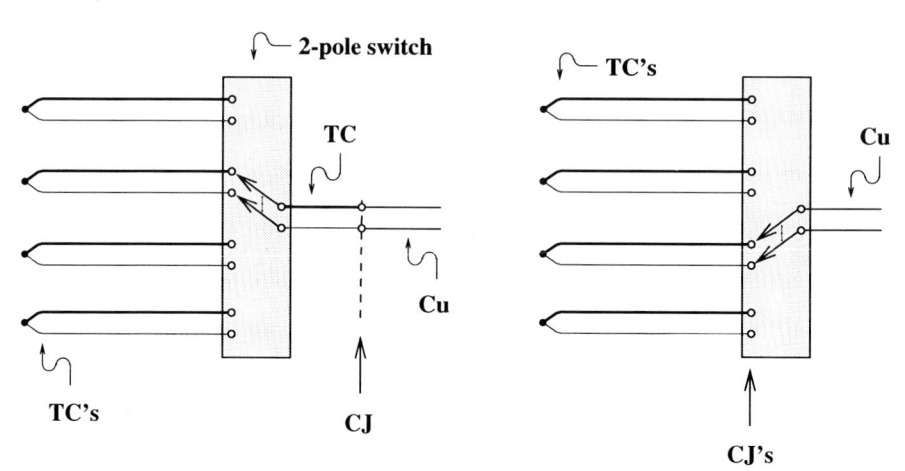

Figure 3.1: Circuits with a switch inserted between thermocouple tips and a common CJ, on the left, and at the thermocouple CJ's, on the right. The thermocouples and their extension leads are indicated by TC, and in the left-hand circuit they must all be of the same type.

In practice, temperature varies along a bank of terminals by ~ 0.1 to $5\,°C$

3.6. SWITCHING

depending on the proximity or otherwise of heat sources, including those within the instrument or switch. In many situations the differences are stable and, as a result, are taken care of by an *in situ* calibration (section 4.10) of each thermocouple. If this is so, or if this level of error is not considered significant, then all is well. Otherwise, they should be measured or avoided by using a different circuit arrangement.

One way of assessing these errors is to check the ACJC adjustment at all input positions. Short lengths of thermocouple wire, cut consecutively from the same reel, are attached to the input terminals and their tips placed together at the same temperature: most easily done in an ice pot. By stepping through the switch positions the different values indicated by the instrument are a measure of the problem. Naturally, the test should be done *in situ* or the switch placed in as severe a thermal environment as would be experienced in its intended application.

It is not enough to simply check all switch positions with copper shorts across the inputs. This checks only for differences in **'thermals'**, thermoelectric signals generated by the switch, which tend to be less of a problem ($< 5\,\mu\mathrm{V}$).

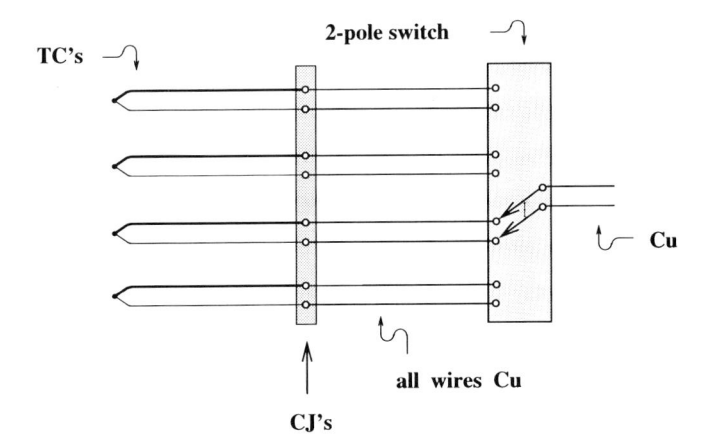

Figure 3.2: Circuit for switching in copper: suitable for thermocouples of mixed type. The CJ's are at the terminals of a uniform-temperature block (see text).

For more accurate work it is better to position the switching function in a copper circuit, as depicted in Figure 3.2, and treat the temperature uniformity across the cold junctions separately. Then, the only thermoelectric performance criterion for the switch is its level of thermals. In the figure, the CJ ends of each thermocouple (or extension lead) meet copper wires at a terminal block, especially designed and located to be uniform in temperature. It is a passive device and being free of moving parts, electronic circuits and transformers it has more chance, than does a driven switch, of being uniform in temperature. A suitable sensor (thermistor or RTD) could be located in the block for the measurement of its temperature (see also Figure 3.5).

To check that the terminal block is uniform in temperature the approach mentioned above for the integrated switch/terminal block could be taken. Alternatively, if a uniformity of $\sim 0.1°C$ or better is desired a differential thermocouple (see section 3.9) arrangement could be used. In which case, the measuring instrument should be disconnected from the switch and the input thermocouples replaced with short lengths of open-circuit thermocouple wire, to avoid electrical shunting of the differential thermocouple and to serve as equivalent heat paths (if significant).

The circuit of Figure 3.2 is suitable for switching thermocouples that differ in type. Their CJ temperatures, being the 'same', are measured at the block and the thermocouple emf's are individually corrected with appropriate values of $V_{CJ \to 0}$, as in equation (3.2). This may be done manually or with a computer/microprocessor. If the thermocouples are all of the same type then the circuit may be modified to take the CJ to the terminals of an instrument, as illustrated in Figure 3.3. Here, a length of thermocouple or extension lead is added to supply emf for the interval from T_0, at the terminal block, to the temperature of the cold junction. Then, the accompanying instrument could be one with an ACJC network and indicate directly in temperature units.

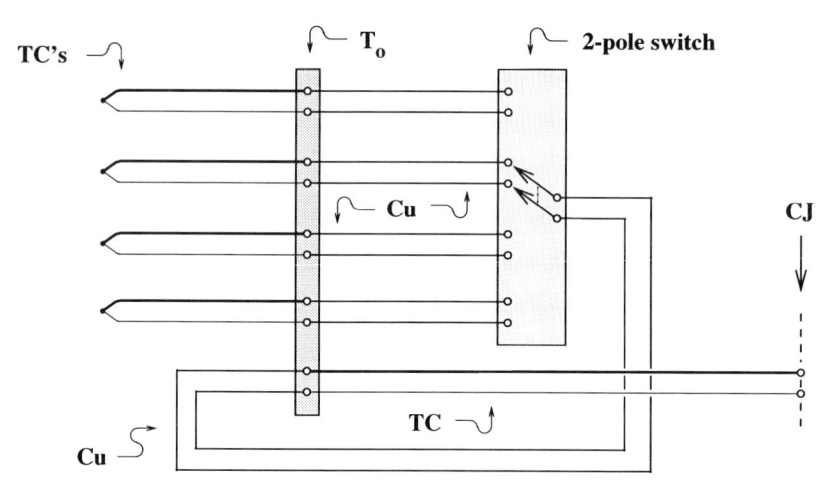

Figure 3.3: Circuit for switching thermocouples in copper: a modification of Figure 3.2, suitable only for thermocouples of the same type.

If an ice pot or an automatic ice point is to be used the circuit could be re-arranged to that shown in Figure 3.4. The emf for the temperature interval T_0, at the terminal block, to 0°C is added electrically and, so, the circuit is

3.6. SWITCHING

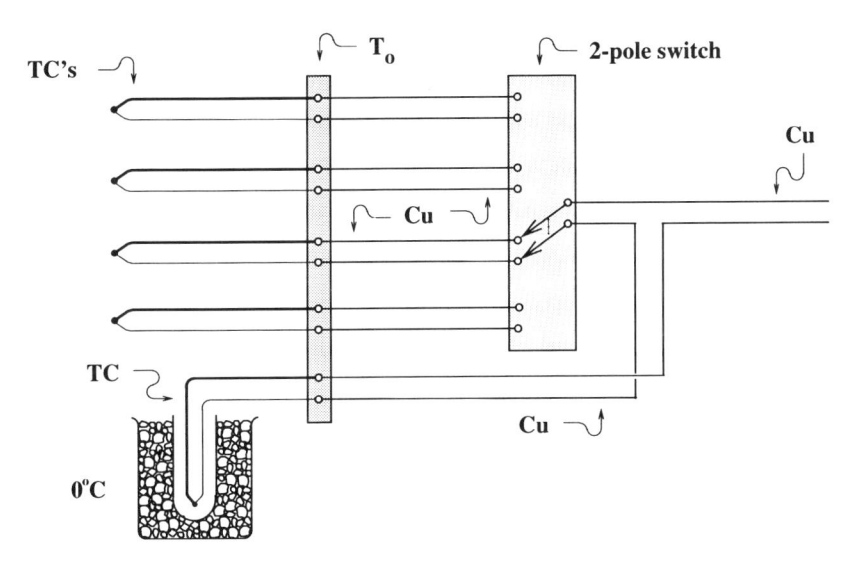

Figure 3.4: Use of an ice pot with thermocouple switching. As shown, the circuit is suitable only for thermocouples of the same type.

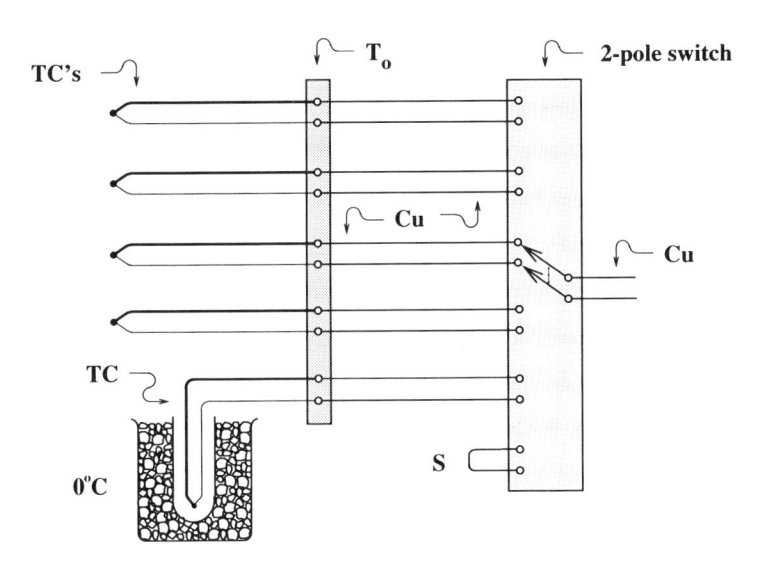

Figure 3.5: An optimal switching arrangement for thermocouples of the same or different type. Shown is a copper short (S) for measuring the system zero and a thermocouple/ice-pot combination for measuring T_0, the temperature of the uniform-temperature block—although other forms of sensor may be used (see text).

suitable only if the thermocouples being switched are of the same type. The configuration may be used with instruments that indicate in emf units or in temperature units (provided the ACJC network can be bypassed, i.e., switched to 'external ACJC').

For thermocouples of mixed type the arrangement in Figure 3.5 could be used. The circuit is suitable only for measurements in emf units and has the added advantage of allowing the system zero to be measured and corrected for using the copper short (S in the figure). The ice pot and its associated thermocouple is used to measure T_0, the temperature of the terminal block, i.e., the CJ temperature of the other thermocouples. Alternatively, T_0 may be measured by another form of sensor, e.g., an RTD embedded in the block. The circuit is optimal since it has the potential for maximum accuracy, is practical and can be used for any thermocouples of the same or of different type.

3.7 Thermocouples in parallel

The average temperature of a region can be determined by measuring the outputs of several thermocouples and forming an average of their temperatures arithmetically. Sometimes it is done by using thermocouples in parallel—to form the average electrically. This is valid provided various conditions are met. The first is that the thermocouples must be electrically isolated from each other, except at the two points where they are connected. Peculiar results are possible, for example, if the tips are placed directly onto an electrically conducting object. The other constraints are best discussed with the aid of Figure 3.6.

The right-hand networks in Figure 3.6 are the equivalent circuits of that on the left, v_j is the emf generated by the jth thermocouple, extending from its tip at temperature T_j to the common CJ at T_0, and r_j is its resistance.

Clearly, the equivalent source impedance of such a network is unchanged if the emf sources are removed. It is therefore given by the parallel combination of all resistances, r_j $(j = 1, \ldots, n)$:

$$r_0 = \frac{1}{\sum_{j=1}^{n} \left(\frac{1}{r_j}\right)}. \tag{3.2}$$

The short-circuit output current, i_0, of the network is the current delivered to a short circuit, if placed across the terminals. Hence:

$$i_0 = \sum_{j=1}^{n} \left(\frac{v_j}{r_j}\right). \tag{3.3}$$

3.7. THERMOCOUPLES IN PARALLEL

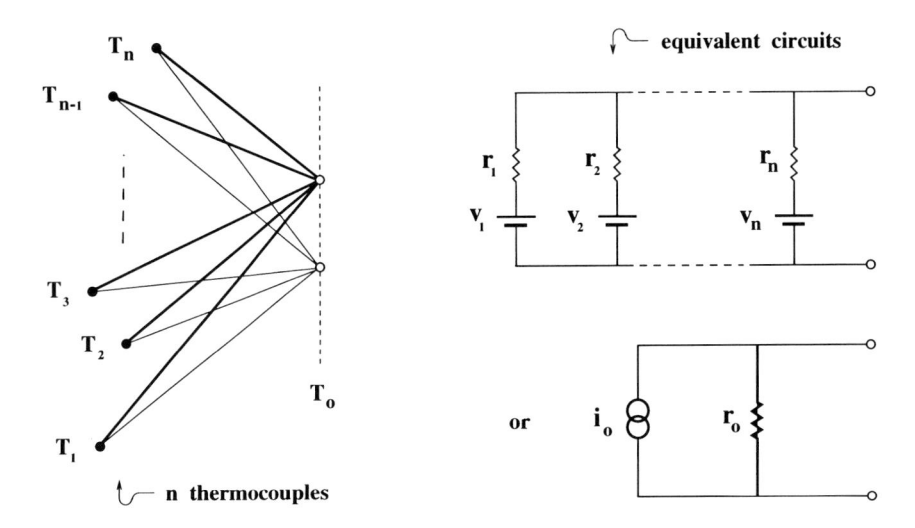

Figure 3.6: Thermocouples connected in parallel with equivalent circuits given on the right.

Therefore, the net emf of the circuit, the signal appearing at the open-circuit terminals, is $v_0 = i_0 \cdot r_0$ or, using equations (3.2) and (3.3),

$$v_0 = \sum_{j=1}^{n} \left(\frac{v_j}{r_j}\right) / \sum_{j=1}^{n} \left(\frac{1}{r_j}\right). \tag{3.4}$$

The output of the parallel combination will equal the average of the thermocouple emf's if all thermocouples have the same resistance, $r_j = r$. Then $\sum(1/r_j) = \sum(1/r) = n/r$, and $\sum(v_j/r_j) = \sum(v_j/r) = (1/r)\sum v_j$. Thus:

$$\begin{aligned} v_0 &= \frac{1}{n}\sum_{j=1}^{n} v_j \\ &= \bar{v}. \end{aligned}$$

It may be difficult to arrange for all r_j to be the same—it is not sufficient for all thermocouples to be of the same type, diameter and length. Differences in temperature profile may result in significant differences in resistance because of the strong temperature dependence of electrical resistivity (see Table 2.4).

The value of temperature obtained from $\bar{v}$, above, would equal the average of the tip temperatures if all the thermocouples have the same calibration and the Seebeck coefficient is independent of temperature over the range involved. Clearly, for a reasonable approximation, all thermocouples should be from the same reel, and the agreement will depend on the extent to which the

Seebeck coefficient varies and how the tip temperatures are distributed over the temperature range involved. For example, the Seebeck coefficient of the type N thermocouple varies with temperature by 12.5% from 200 to 400°C. So the use of 5 thermocouples in parallel to obtain an average temperature in this temperature range would yield an error of up to 2°C. The maximum error is larger for the range 0 to 200°C, where it is 4.3°C, because here the Seebeck coefficient varies by 27%. Even larger errors occur if type T thermocouples are used instead of type N: up to 5.8°C for this range.

The optimum choice of thermocouple material for 'parallel' applications that involve large temperature differences is the type K. Its Seebeck coefficient varies the least with temperature (see Table 2.2 on page 27). Thus, for 5 type K thermocouples in the range 0 to 200°C the error in a calculated average would not exceed 0.5°C: smaller by an order of magnitude than the error for type N or T.

Another point worth noting is that inhomogeneity in Seebeck coefficient represents an uncertainty (see page 172) of about 0.1% in each thermocouple signal. The effect of this on the calculated average is smaller, it is $0.1/\sqrt{n}\%$. This improvement does not apply, however, to changes in inhomogeneity that occur equally in each thermocouple during their use in the parallel configuration. Such changes are systematic (section 5.3) and are characteristic of the type of thermocouple material and of the temperature range.

In practice, the CJ could be transferred from the common terminals at temperature T_0 (Figure 3.6) to the input terminals of an instrument with ACJC using a suitable extension lead.

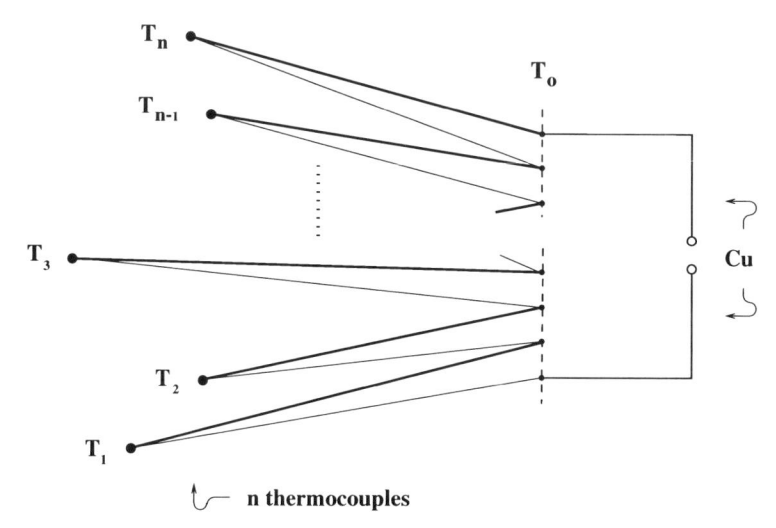

Figure 3.7: Thermocouples connected in series addition.

3.8 Thermocouples in series

Placing thermocouples in 'series addition', as shown in Figure 3.7, allows an average tip temperature to be obtained with the benefit of increased sensitivity. The thermocouples should have the same value of CJ temperature (T_0). Clearly, the resistance of the network is the sum of the individual thermocouple resistances and the total output is the sum of the developed emf's. Hence, if the total output is divided by the number of thermocouples a true average of thermocouple emf is obtained. Unlike the parallel network (section 3.7) there is no need for the thermocouple resistances to be equal. Nevertheless, the remarks made about the average emf—as not necessarily corresponding to the average temperature—still applies.

Notice, that since the effective Seebeck coefficient is greater, an extension lead—to extend the circuit beyond T_0—and temperature-indicating instruments cannot be used.

As with thermocouples in parallel, if significant electrical leakage exists between the thermocouples a nonsensical result may occur.

3.9 Differential thermocouples

In previous sections, thermoelectric circuits were treated in terms of wire pairs with the wires of each pair existing side by side and operating over the same temperature interval. Nevertheless, it is also useful to consider thermocouples in a three-wire form, as illustrated in Figure 3.8. In some ways, this representation follows more easily from the discussion on Seebeck emf in section 1.3. There, it was shown that the Seebeck emf in a length of wire could be used as a measure of the temperature at one of its ends, at T_1 for example, if that at the other end, T_2, was known. To measure the signal two connecting wires may be added, and these of course would contribute emf of their own. The resultant configuration is shown in Figure 3.8, on the left, together with its thermoelectric equivalent, on the right. Here, it is represented as a two-wire thermocouple, spanning the temperature interval T_1 to T_2, connected to a pair of wires generating no emf for the interval T_2 to T_3. This is provided that the two connecting wires are of the same material, metal 2, and are terminated at the same temperature, T_3. The output of the three-wire thermocouple is not a function of T_3.

When constructed in the three-wire form (Figure 3.8) a thermocouple is referred to as a **differential thermocouple**, especially if the mean Seebeck coefficient, $\bar{S}$, for temperatures from T_1 to T_2, can be estimated without

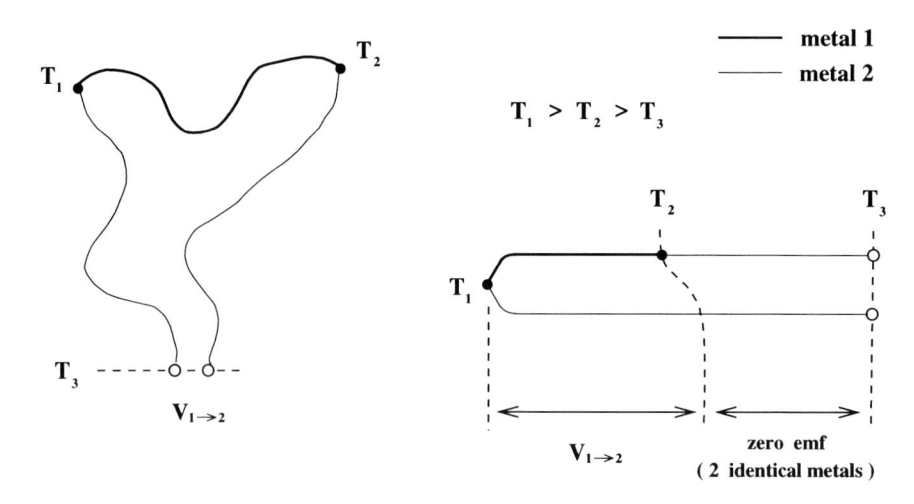

Figure 3.8: A thermocouple presented in three-wire form, as it would be used, on the left, and re-drawn on the right to express its signal in terms of wire pairs.

knowing either T_1 or T_2. Then, we have

$$V_{1 \to 2} = \bar{S}(T_1 - T_2). \tag{3.5}$$

The signal, $V_{1 \to 2}$, is a direct indication of the temperature difference $(T_1 - T_2)$.

For equation (3.5) to represent a practical method for measuring temperature differences directly, $\bar{S}$ needs to be estimated without accurate knowledge of either T_1 or T_2. This in turn requires that a rough estimate of the mean temperature be made and that the difference $(T_1 - T_2)$ be restricted. These are not difficult requirements. For example, the Seebeck coefficient of type S thermocouples is within $\pm 5\%$ of $10.5\,\mu\text{V}\,\text{K}^{-1}$ from 550 to 850°C. Thus, if T_1 and T_2 were known to be within this relatively large temperature range the value of $V_{1 \to 2}$, from equation (3.5) with $\bar{S} = 10.5\,\mu\text{V}\,\text{K}^{-1}$, would represent $(T_1 - T_2)$ to better than 5%. Similarly, a type K thermocouple may be used to measure differences to better than 4% for temperatures between 0 and 950°C, where $\bar{S} = 41.0\,\mu\text{V}\,\text{K}^{-1}$, and to better than 1% if the temperature is known to be between 335 and 715°C, say, where $\bar{S} = 42.2\,\mu\text{V}\,\text{K}^{-1}$.

The accuracy of differential thermocouples is not usually limited by the knowledge of $\bar{S}$. Of greater difficulty is the requirement that the net emf developed from T_2 to T_3 be zero. This is particularly so if $(T_2 - T_3)$ is large.

Consider a typical application for differential thermocouples: the measurement of small temperature differences, of less than 1°C, in an oil bath at a temperature 200°C above ambient. The desired signal derives from a temperature difference of less than 1°C, yet each wire passing out through the oil/air interface produces emf for a 200°C interval. The emf's in these

3.9. DIFFERENTIAL THERMOCOUPLES

connecting wires must match each other to better than 0.05% to affect the measured difference by less than 10%. This is best done by choosing a homogeneous material, such as (potentially) copper or platinum, for metal 2 (Figure 3.8). An alternative approach is to hold the mismatch error constant, by avoiding significant changes in the temperature profile along the connecting wires. Then, the error may be measured as the system zero and corrected for—as done in the following example.

One method of probing for temperature differences in an oil or water bath is to use flexible wires attached to glass rods, for moving and positioning the two junctions. To greatly reduce mismatch and inhomogeneity effects in the connecting leads the wires may be clamped where they enter the bath at the liquid/air interface. Then a major part of their emf is fixed while the two junctions are moved about. Any calibration difference between the two wires would be a fixed contribution to each measurement made with the set-up, as would be (in principle at least) the instrument zero and circuit thermals. The combined effect of these fixed (systematic) errors is corrected for by subtracting the system zero—the emf measured with the tips held together ($T_1 = T_2$)— from the emf measured with the tips apart. In this way mismatch errors may be reduced to $\sim 0.01\,°C$ or better. The instrument zero and circuit thermals would need to be stable to an equivalent level.

In summary, consider the following points:

- In percentage terms, the accuracy required for the Seebeck coefficient of a differential thermocouple is roughly that required for $(T_1 - T_2)$. This in turn dictates how well the average temperature is to be estimated. If, for example, the Seebeck coefficient for a type T thermocouple is required to within $\pm 5\%$ the temperature needs estimating to $\pm 30\,°C$ at $100\,°C$ or $\pm 50\,°C$ near $200\,°C$. On the other hand, the temperature hardly needs estimation for a type K differential thermocouple, the coefficient is $41.0\,\mu V\,K^{-1}$ to $\pm 5\%$ over the range, 0 to $1000\,°C$.

- The Seebeck coefficient of standard-grade wire is probably within $\pm 2\%$ of that indicated by the reference tables. For a more reliable estimate of Seebeck coefficient either premium-grade wire ($\sim \pm 1\%$) should be used or the coefficient determined from a calibration over the temperature range of interest.

- The emf's developed from T_2 to T_3 by the two connecting wires, of metal 2 in Figure 3.8, should cancel. If the thermocouple type is T the Constantan wire should be chosen as metal 1 and the potentially homogeneous copper wire would then serve excellently as metal 2. Similarly, platinum is preferable to the platinum-rhodium alloy as metal 2 if type S or type R thermocouples are used.

- The open ends of the thermocouple ought to be at the same temperature, T_3. To ensure this, it may prove necessary to locate them in an ice pot or some other uniform enclosure. If the differential thermocouple is a type T, with copper as metal 2, the problem does not arise because low-level voltage measuring instruments have input circuits of copper or of metals with a similar Seebeck coefficient.

- Differential thermocouples supply estimates of temperature difference. If the temperature at any particular site is also required it will need to be measured as a separate exercise.

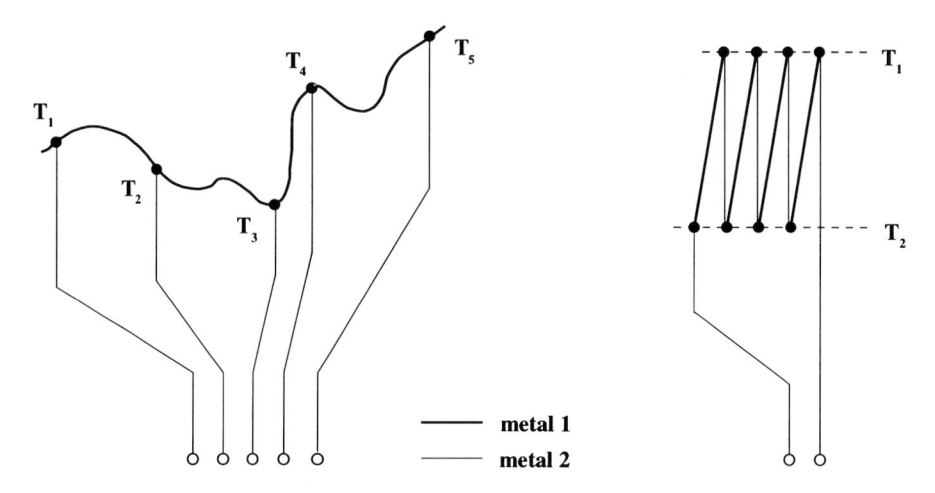

Figure 3.9: Differential thermocouples in sequence, on the left, and in series, on the right.

Figure 3.9 illustrates two extensions to the basic format of a differential thermocouple. The left-hand circuit shows the ease with which the temperature difference between any pair of several sites can be determined. Simply run a length of Constantan wire, for example, from site to site in any order and then run separate lengths of copper from each site to a switch. The measured output from any two copper wires is a direct indication of the temperature difference between the junctions to which they connect. This arrangement may be more convenient to use than the movable version described above, and measurements can be taken more rapidly. There is then a greater need for the return wires to have matching, and hence cancelling, emf's. Also, the switch will introduce unwanted Seebeck emf's ('thermals') of its own. Low-thermal rotary switches may have thermals as small as $\sim 0.1\,\mu$V and relays are available with low ($< 1\,\mu$V) to moderate ($< 5\,\mu$V) levels of thermals.

The right-hand circuit of Figure 3.9 illustrates a means of increasing the

3.10. GAS TEMPERATURE

sensitivity of differential thermocouples. For example, if 10 thermocouples, each operating between T_1 and T_2, were joined in series addition the effective Seebeck coefficient of the network would be 10 times greater than that of any one thermocouple. Such a network is sometimes called a thermopile.

3.10 The measurement of gas temperature

The measurement of gas temperature deserves special treatment. The heat transfer (convection) from a gas to a temperature probe, placed in the gas for its measurement, tends to be smaller than the radiative interchange between the probe and surrounding surfaces. Indeed, the thermal state of the probe will often be influenced more by the temperature of the wall than by that of the gas. Large errors should be expected unless the entire region is uniform in temperature—being nominally-uniform is not enough (see pages 109-109).

Consider a gas within an enclosed region, such as a flue pipe or an oven. For convenience, assume the gas is hotter than that of the wall (the alternative assumption will affect only the sign of any calculated outcome). If a thermocouple is placed at the site of interest its temperature will rise until at equilibrium the heat gained from the gas, by convection, balances that lost by radiation to the walls and by conduction along the thermocouple leads. For this discussion let us assume the geometry and placement of the thermocouple is such as to avoid significant conduction (near the tip)—usually a relatively simple matter to arrange. So, for calculation purposes the thermocouple will be taken as a simple cylinder (without leads) having a length equal to that of the uniform-temperature region near its tip—as shown in Figure 3.10.

In the figure, the heat transfer mechanisms are represented by thermal resistances, an electrical-analogue approach discussed elsewhere [103, 104]. The method is particularly useful in qualitative discussions. For example, if any net heat flow, Q, exists via the thermocouple, as shown, as it must if the gas and wall temperatures differ, the temperature of the thermocouple is dictated by the relative magnitudes of the two resistances R_{rad} and R_{conv}. If the resistances are equal, for example, as they would be for natural convection near 200°C (see below), the temperature of the thermocouple will be midway between those of the gas and wall. To achieve an accurate measurement of gas temperature, the flow Q needs to be small (large R_{rad}: reduced radiation) and/or the resistance, R_{conv}, reduced (improved convective transfer).

The heat gained by the thermocouple, at temperature T_{tc}, is that convected by the gas, at T_g:

$$Q_C = hA_{tc}(T_g - T_{tc}), \tag{3.6}$$

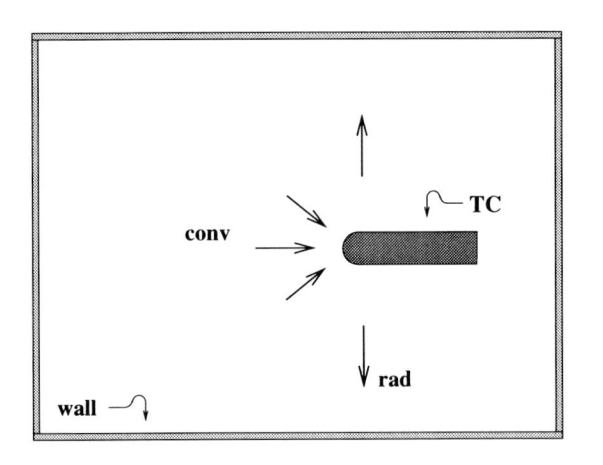

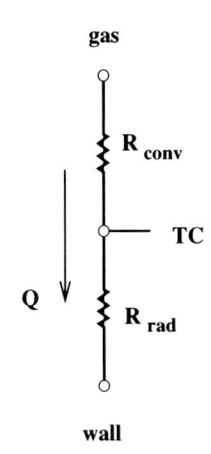

Figure 3.10: Thermocouple (TC) probe within a chamber: temperature of TC is governed by thermal resistance (R) to the gas (convection), locally assumed hotter than the wall, thermal resistance to the wall (radiation) and the heat flow (Q).

where h is the convective (or film) heat transfer coefficient and A_{tc} is the effective surface area of the thermocouple probe. This equation understates the complexity of the convection mechanism and as a result the coefficient h, defined in this way, depends on the shape, roughness and inclination of the surface, as may be expected, but also on the difference, $T_g - T_{tc}$ and on whether convection is natural or forced.

The radiative heat transfer from the thermocouple to the walls, assumed at a uniform temperature of T_w, is [104]

$$Q_R = \epsilon_* \, \sigma \, A_{tc} (T_{tc}^4 - T_w^4),$$

where σ is the Stefan-Boltzmann constant, 5.67×10^{-8} W m^{-2} K^{-4}, and the parameter ϵ_* depends on the emissivities of the thermocouple and the wall, ϵ_{tc} and ϵ_w, via the relationship

$$\frac{1}{\epsilon_*} = \frac{1}{\epsilon_{tc}} + \frac{A_{tc}}{A_w}\left(\frac{1}{\epsilon_w} - 1\right),$$

which is a good approximation for a completely convex object placed within a completely concave surface [105]. Assuming the surface area of the thermocouple is much smaller than the inner surface of the wall, this reduces to

$$\epsilon_* = \epsilon_{tc}, \quad \text{and so}$$

$$Q_R = \epsilon_{tc} \, \sigma \, A_{tc}(T_{tc}^4 - T_w^4). \tag{3.7}$$

3.10. GAS TEMPERATURE

At equilibrium we have $Q_C = Q_R$ and, from equations (3.6) and (3.7)

$$T_g - T_{tc} = \frac{\epsilon_{tc}}{h}\sigma(T_{tc}^4 - T_w^4). \tag{3.8}$$

As expected, the measured gas temperature, i.e., that of the thermocouple, T_{tc}, is lower than the true temperature, T_g, and the error is

$$\Delta = T_{tc} - T_g.$$

Once a measurement T_{tc} is made and T_w is known, then the gas temperature may be calculated from equation (3.8) quite simply, while noting that temperatures are absolute (kelvin). On the other hand, for design purposes it is useful to determine the likely error, $T_{tc} - T_g$, for various choices for T_g and T_w. Then, because T_{tc} is not separable in the equation, an iterative procedure is required, as follows.

1. Select a first guess for T_{tc}, such as $T_{tc1} = (T_g + T_w)/2$, and

2. calculate the gas temperature, $T_{g,calc}$, that corresponds to this estimate, from equation (3.8).

3. $T_{g,calc}$ is higher than T_g by $(T_{g,calc} - T_g)$, so reduce the estimate of T_{tc} to

$$T_{tc2} = T_{tc1} - (T_{g,calc} - T_g)\left(\frac{\delta T_{tc}}{\delta T_g}\right)$$

where, from equation (3.8), we have

$$\frac{\delta T_{tc}}{\delta T_g} = 1 + 4\frac{\epsilon_{tc}}{h}\sigma T_{tc1}^3.$$

4. Return to step 2, above, and repeat until $(T_{g,calc} - T_g) < 0.01°C$, or whatever, i.e., until the calculated outcome is considered close enough.

Some values obtained in this way are given in Table 3.1.

Clearly, it is difficult to measure gas temperatures by placing a simple temperature probe in the gas. The problem is greatest when attempting to measure a hot slowly moving gas (low h) in a duct or chamber whose wall is considerably colder than the gas. As for reducing the error, once again consider equation (3.8). The error is proportional to ϵ_{tc}/h and is smaller if ϵ_{tc} is effectively decreased and/or h is increased. Some suggestions for making such an improvement are given below.

- The radiative loss from the thermocouple to the wall may be reduced by surrounding it with a radiation shield, as discussed in detail below.

Table 3.1: Values of Δ, the error in measured gas temperature using an unshielded probe, for various gas (T_g) and wall temperatures (T_w). The following were assumed: $\epsilon_{tc} = 0.4$, $h = 10\,\mathrm{W\,m^{-2}\,K^{-1}}$ for natural convection and $h = 100$ for forced.

T_g (°C)	Δ (°C)				
	$T_w = 100\,°\mathrm{C}$	$T_w = 200\,°\mathrm{C}$	$T_w = 300\,°\mathrm{C}$	$T_w = 400\,°\mathrm{C}$	$T_w = 500\,°\mathrm{C}$
Natural					
110	−3.3				
210	−42	−4.9			
310	−90	−58	−6.3		
410	−146	−118	−72	−7.4	
510	−207	−182	−141	−82	−8.1
610	−272	−250	−213	−159	−90
1000	−556	−540	−513	−471	−416
Forced					
610	−87	−82	−73	−58	−36
1000	−247	−244	−237	−226	−209

- In many cases a single radiation shield will not be sufficient. Then, a series of coaxial shields may be used, although such an approach may be impractical. For example, in an attempt [106] to measure gas temperatures in the range 140 to 400°C eight concentric shields were found necessary.

- Rather than use a multiple-shield design, the loss from a single radiation shield may be reduced by surrounding it with a porous ceramic sleeve, which represents a significant series impedance (relatively poor thermal conductivity), or by adding a heater winding (if the gas is hotter than the walls). In the latter case, a second thermocouple placed along the shield indicates when its temperature matches that of the gas thermocouple. Then, no loss to the shield can occur and the measured gas temperature is correct.

- A common approach is to use a suction pyrometer [105, 107]. It effectively decreases ϵ_{tc}, using at least one radiation shield, and increases h by drawing gas through the innermost shield, over the thermocouple, using a suction pump. The method involves some trial and error.

Usually, gas temperatures are more easily measured when convection is strong (high gas flow). On the other hand, the measurement is more difficult if the gas velocity is too large, such as in the exhaust of an aircraft engine. Here, the gas on coming to rest when striking the thermocouple, loses energy

3.10. GAS TEMPERATURE

to it, so causing an increase in temperature. The increase is [105]:

$$\Delta T \approx c v^2,$$

where v is the gas velocity in m/s and the constant c varies from 0.005 at 0°C to 0.0042 at 1000°C. For example, a 1°C rise would occur at about 15 m/s.

Use of a radiation shield

As a guide to what may be achieved in using radiation shields consider the following simple case. Ignore the thermocouple leads, as before (no conduction loss), and assume the probe is a closed cylinder with a surface area, A_{tc}. Surround the thermocouple with a relatively long, cylindrical radiation shield, as depicted in Figure 3.11, long enough to ignore the radiative loss through its open ends.

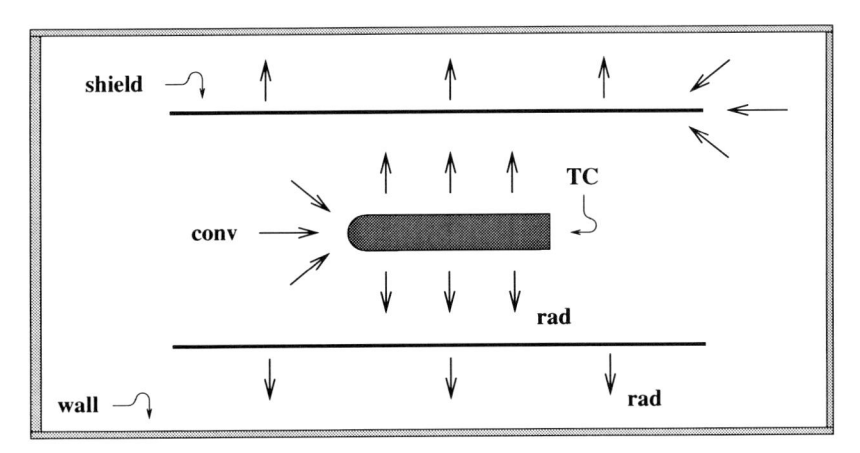

Figure 3.11: Thermocouple (TC) probe within a radiation shield: temperature of TC is governed by convection from gas and radiation to the shield, which in turn is heated by the gas and cooled by the wall.

The electrical equivalent circuit for this arrangement is given in Figure 3.12. Under steady state conditions the temperatures of the thermocouple and shield are determined by those of the gas and wall as well as the thermal resistances shown in the figure.

The heat balance for the thermocouple is similar to that given by equation (3.8), with T_{sh} as the shield temperature:

$$T_g - T_{tc} = \frac{\epsilon_{tc}}{h_{tc}} \sigma (T_{tc}^4 - T_{sh}^4). \tag{3.9}$$

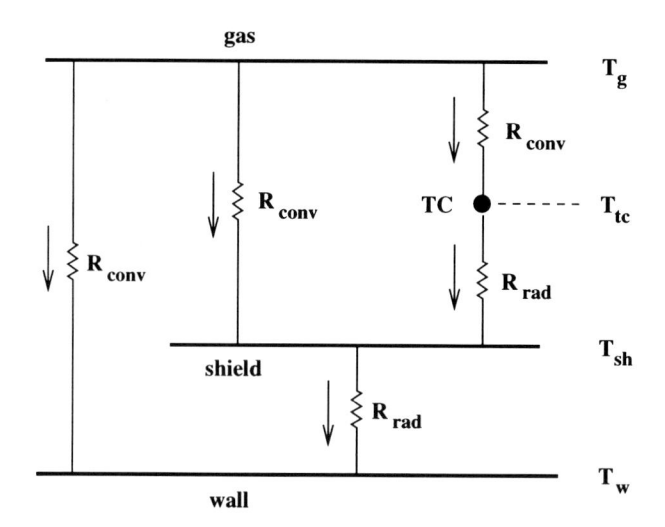

Figure 3.12: Electrical equivalent circuit for Figure 3.11 (a shielded thermocouple (TC)), showing thermal resistances due to convection (conv) and radiation (rad) and the resultant flow paths and temperatures (T).

Similarly, the heat balance for the radiation shield, having two surfaces heated by convection (each surface of area A_{sh}) and receiving radiation from the thermocouple, is

$$2(T_g - T_{sh}) = \frac{\epsilon_{sh}}{h_{sh}}\sigma(T_{sh}^4 - T_w^4) - \frac{\epsilon_{tc}}{h_{tc}}\sigma(T_{tc}^4 - T_{sh}^4)\frac{A_{tc}}{A_{sh}}. \qquad (3.10)$$

The term on the right is the radiant contribution from the thermocouple and, being relatively small, may be ignored. Also, for convenience, make the reasonable assumption that the emissivities and the heat transfer coefficients for thermocouple and shield are the same. Then, equations (3.9) and (3.10) become

$$T_g - T_{tc} = \frac{\epsilon}{h}\sigma(T_{tc}^4 - T_{sh}^4) \quad \text{and} \qquad (3.11)$$

$$T_g - T_{sh} = \frac{\epsilon}{2h}\sigma(T_{sh}^4 - T_w^4). \qquad (3.12)$$

Notice, on comparing equations (3.8) and (3.12), that the shield settles on a temperature equal to that of an unshielded thermocouple having an emissivity of $\epsilon/2$.

Using an iterative procedure, like that described above for Table 3.1, the error in gas measurement, $\Delta = T_{tc} - T_g$, was obtained for various gas and wall temperatures and appears in Table 3.2.

3.10. GAS TEMPERATURE

Table 3.2: Values of Δ, the error in gas temperature (T_g), when a radiation shield is interposed as shown in Figure 3.11. The following were assumed: $\epsilon = 0.4$, $h = 10\,\mathrm{W\,m^{-2}\,K^{-1}}$ for natural convection and $h = 100$ for forced.

T_g (°C)	Δ (°C)				
	$T_w = 100\,°C$	$T_w = 200\,°C$	$T_w = 300\,°C$	$T_w = 400\,°C$	$T_w = 500\,°C$
Natural					
110	−0.7				
210	−13	−1.7			
310	−36	−25	−3.0		
410	−69	−58	−38	−4.3	
510	−111	−101	−82	−51	−5.5
610	−159	−150	−133	−105	−63
1000	−392	−386	−373	−352	−321
Forced					
610	−19	−18	−16	−13	−8
1000	−96	−95	−93	−90	−84

Temperature in ovens and furnaces

Thermocouples are often used to measure the spatial distribution of temperature throughout heated enclosures, such as ovens and furnaces (Chapter 6). Thermocouples are placed at various positions throughout the region and the temperatures so obtained are taken as values of gas temperature. It is further assumed that the measured distribution applies equally to the specimens or components subsequently treated in the oven. These assumptions will now be examined.

Just as temperature varies throughout the gas, i.e., the internal space of the enclosure, it also varies over its walls, by at least as much. Indeed, the convection process in the gas is driven, at least in part, by the gross variations in temperature over the walls. By 'walls' I mean the surfaces that surround the working space of the enclosure and includes any exposed heaters, if present.

At any one site within the working space a probe will be influenced by the local temperature of the gas and by an effective, mean wall temperature, as seen from the perspective of the probe. As a result, the probe temperature will lie somewhere between the two, as may be seen in the table on page 106.

With an oven operating near ambient temperature, a probe will have a temperature closer to that of the gas than to the wall. As the (mean) oven temperature is increased the probe temperature will move progressively closer to that of the wall, as can be seen in Table 3.3. Here, the data were obtained by solving equation (3.8) for small values of $(T_g - T_w)$ and, as before, the temperature probe was treated as a small, uniform-temperature cylinder with no conducting link to the walls. In the table, a value of 5 W m^{-2} K^{-1} was used for the film coefficient h when representing gravity/natural convection, a value most likely to apply to the nominally-uniform conditions within an oven.

It is instructive to calculate that oven temperature at which radiation and convection are equally significant in establishing the temperature of a probe, or of a small object being treated in the oven. When this condition applies the temperature of a thermocouple will lie midway between that of the gas and the wall. Since T_w and T_{tc} are of a similar magnitude (in a nominally-uniform oven or furnace) we can approximate equation (3.8) as

$$T_g - T_{tc} \approx 4\frac{\epsilon_{tc}}{h}\sigma \bar{T}^3 (T_{tc} - T_w),$$

where $\bar{T}$ is the average: $(T_{tc} + T_w)/2$. When T_{tc} is midway between T_g and T_w, we have $(T_g - T_{tc}) = (T_{tc} - T_w)$ and

$$\bar{T}^3_{mid} = \frac{h}{4\epsilon_{tc}\sigma}. \tag{3.13}$$

For the purposes of this exercise, we can take $\bar{T}_{mid}$ as the oven or furnace temperature at which radiation and convection are of equal significance.

Values of $\bar{T}_{mid}$ are given in Table 3.4. For example, $\bar{T}_{mid}$ is 107°C for $h = 5$ and $\epsilon_{tc} = 0.4$. In other words, in a gravity-convection oven operating near 100°C convection and radiation have an equal influence on the temperature of a probe, or of any small object ($\epsilon = 0.4$) sitting free of the walls, and above this temperature radiation will dominate.

Clearly, gas temperature is typically **not** measured in ovens and furnaces! The temperature measured at any one site is dictated mainly by the radiative interchange with the surrounding surfaces, and its value depends on the emissivity of the probe and on the view factor of each significant hot and cold spot of the walls. Once the enclosure has been tested with temperature probes the measured temperature distribution will be presumed to apply to a variety of objects subsequently placed in the oven (or furnace) for thermal conditioning (heat treatment). This may be close to the truth if the temperature over the surface of the walls is uniform and if the emissivities of the objects being conditioned equal that of the test thermocouples.

3.10. GAS TEMPERATURE

Table 3.3: Difference in temperature, $(T_g - T_{tc})$, between that of a probe and that of the gas in a nominally-uniform-temperature oven, expressed as a percentage of the overall difference $(T_g - T_w)$. Values are given for various oven temperatures, with $\epsilon_{tc} = 0.4$, for natural ($h = 5$) and forced convection ($h = 100$) and a probe without radiation shielding.

Temperature (°C)	$(T_g - T_{tc})/(T_g - T_w)$ (%)	
	$h = 5$	$h = 100$
20	31	2
50	38	3
100	49	5
150	58	6
200	66	9
300	77	15
400	85	22
500	89	30
1000	97	65

Table 3.4: Values of $\bar{T}_{mid}$, representing the nominal oven/furnace temperature at which the temperature of a probe lies midway between that of the gas and that of the wall, given as a function of ϵ_{tc} and h (W m^{-2} K^{-1}).

h	$\bar{T}_{mid}$ (°C)		
	$\epsilon_{tc} = 0.2$	$\epsilon_{tc} = 0.4$	$\epsilon_{tc} = 0.6$
1	7	−51	−79
2	80	7	−28
5	206	107	59
10	331	206	146
20	488	331	254
50	760	546	443
100	1028	760	629
200	1366	1028	863

Chapter 4

The Calibration of Thermocouples

4.1 Introduction

Do thermocouples need calibration? The answer is that some do and those that do not are not calibrated because the manufacturer's tolerances (section 2.2.1) for thermocouples and instruments are regarded as adequate safeguards against the inaccuracy of measurement. A thermocouple will most likely begin its service within specification and remain so for several days, the time depending on the temperature of use. The drift in calibration will tend to increase with temperature and may be minimised by using a fully-protected Pt-based thermocouple or a Ni-based thermocouple in the ID-MIMS format (section 2.6.5). For applications where a larger drift can be tolerated base-metal thermocouples in either a conventional MIMS design (section 2.6.4) or as heavy-gauge bare wires (section 2.6.2) may be suitable.

On the other hand, many industrial processes require a better knowledge of temperature. Then, the thermocouple, or a sample from the same batch, should be calibrated to indicate its initial departure from the nominal emf-temperature relationship. Furthermore, its inevitable drift from this state could be kept manageable by fixing it in position, with respect to the temperature gradient zone, monitoring the drift by periodic *in-situ* calibrations and making appropriate corrections. This simple approach enables the manufacturing process to be kept close to its required temperature, and any change in the control equipment or the process variables can be accommodated with a minimum of product wastage. Knowledge of the calibration of key components in a control process, with control settings corrected accordingly, is the most efficient safeguard in quality control.

Unfortunately, many operators have followed a tradition of doing without calibration, even in situations where knowledge of temperature is critical. Instead, temperature is often taken as a somewhat arbitrary parameter that is used only to reproduce conditions on a day to day basis. In the long term, control settings are regularly tweaked to maintain a satisfactory product, as judged by measurements taken at the last stage of production. This is an inefficient and costly alternative that has its historical roots in two perceptions [108]. Firstly, that a thermocouple is a crude device with poor stability, an understandable view when you consider many control thermocouples— lengths of 'fencing wire' mounted in beads! Secondly, that thermocouple emf is developed in its welded tip (see section 1.3). Both perceptions have generally led to a mistrust in the device and to inappropriate and thus inefficient practices.

Naturally, thermocouples used to measure temperatures at critical sites within an industrial process, or to calibrate its control sensor, require calibration. So also do those used for research and development. In most such cases calibration needs to be done at a higher level of accuracy than is required by process monitoring thermocouples.

Furthermore, every calibration of temperature sensors should be traceable to ITS–90. This implies the existence of a hierarchy of calibration laboratories [109] between the end user and a national standards laboratory, with a progressive increase in calibration accuracy from one to the other. At each level of the hierarchy calibrated thermocouples may be involved.

4.2 Calibration

The word 'calibrate' is derived from 'calibre' and its original meaning was 'to measure the calibre of' something. Over the years it has been used by manufacturers to mean applying graduation marks (to pipettes and thermometers, for example) after first measuring the appropriate parameter (volume and temperature, respectively). In general, calibration may be defined as the measurement of an individual characteristic of a device, and, in particular, of how this value differs from some stated, nominal value.

In other words, calibration may be regarded as the measurement of that systematic error (see section 5.3) that characterises a device, provided the error is stable enough for calibration to be worthwhile. Calibration, as it applies to instruments, is often taken to mean 'an adjustment' to bring the behaviour back within specification. On the other hand, thermocouples, like length tapes and potentiometers, are not user adjustable, and the calibration of a thermocouple will involve measurements of its particular emf-temperature relationship and should indicate how it differs from the nominal relationship,

4.2. CALIBRATION

as given in the standard reference tables (section 1.4). Thus, it is not the thermocouple that is 'adjusted' but its emf (or the equivalent temperature), arithmetically, by applying a correction.

A calibration is essentially a measurement and, like all measurements, its result should consist of two parts—a correction (arithmetic or manual), to adjust the measured parameter to its nominal value, and the calibration uncertainty, a measure of how wrong the correction could be. In the case of a manual adjustment (instruments etc.), the uncertainty of adjustment (setting the error to zero) may be taken as the manufacturer's specified tolerance. The magnitude of each correction depends on the particular set of conditions that applied during the test and the calibration uncertainty depends on the technique, the equipment, the level of expertise and to what extent the calibration result is to refer to future use. Is the calibration result valid only at the time of test or does it also refer to a defined 'period of validity'?

Often, instruments and thermocouples arrive at a calibration laboratory with a request that may be paraphrased as: "thermocouple, 1 off, please calibrate". Time is then wasted in tracking down the intended user of the thermocouple to ascertain the required level of accuracy and the temperature range for the calibration This is usually rewarded with information of the kind: "don't know" or "you should know the temperature range for that sort of thing!". A type S thermocouple can operate to a least 1700°C. For the CSIRO National Measurement Laboratory (NML) to calibrate it up to this temperature would entail three times the cost of a calibration limited to 1100°C. Moreover, different levels of calibration uncertainty may be assigned depending on the assumed use for the thermocouple.

Regardless of whether a thermocouple is calibrated by the user or sent to an independent laboratory for its calibration, some knowledge of thermocouple behaviour is required. Reasonable decisions on how the thermocouple should be treated in calibration and in its subsequent use can be made only with such knowledge. The calibration laboratory may not have available the best or most appropriate option and the contact person may be ignorant of the principles, as outlined in this chapter.

4.2.1 Points to consider in thermocouple calibrations

Here I deal with general questions on calibration, all of which arise because the thermocouple is not a well-behaved point source of emf, as is often assumed. The calibration of reference standards is discussed in the next section (the NML procedure, on page 122) and the use of reference standards to calibrate other thermocouples, in section 4.4. So let us begin with a number of aspects of thermocouple behaviour and the consequences for calibration.

- The thermocouple generates a low-level DC signal, which depends on the temperatures of both the CJ and the tip.

- Seebeck emf is distributed along the conductors according to their thermoelectric signatures and the temperature profile experienced at that time.

- In general, temperature varies from point to point along a thermocouple. So, any change in Seebeck coefficient that occurs during use will be non-uniform.

- If a calibration is conducted at one immersion depth only, as is usually the case, the measured emf is derived from the section of thermocouple in the temperature gradient region, extending maybe ~ 200 mm. The calibration data may not be applicable to other sections that may generate the signal in a subsequent use—an assumption of homogeneity is required.

- Implicit in a request for re-calibration is the possibility of change and, since change in the Seebeck coefficient in not uniform, the main function of re-calibration should be to scan the thermocouple length and characterise the change (see section 4.4). Nevertheless, a scan is not required for those thermocouples used and calibrated at the same position in a furnace (section 4.10).

Calibration can also be looked at in terms of its purpose: the use to which a calibration is put. This should be defined so that the most suitable calibration can be arranged. Calibrations fall into four categories, when classified according to purpose.

1. An acceptance test, or the equivalent, for thermocouples purchased to a particular manufacturing tolerance. Often, this applies to coils of thermocouple cable/wire and the short test lengths cut from the coil for calibration may never be used again.

2. A batch calibration: here a small number of thermocouples, at least two, are selected to represent a batch or coil and are calibrated, and possibly discarded. Their mean calibration is taken as the calibration of all thermocouples from the batch, or coil, and the difference between their calibrations is a measure of the variation within the batch, or along the coil.

3. The calibration of a thermocouple to be used at a fixed location. Its initial emf-temperature relationship is all that is required and any drift that may then occur is considered separately. The drift may be ignored,

4.2. CALIBRATION

if the period of use is to be short or the temperatures low, otherwise it may be determined by periodic *in-situ* calibrations.

4. The calibration of reference standards or thermocouples to be used as working-standards or as test probes, destined for several short-term uses, possibly in different furnaces and at differing immersion depths. Here, the main criterion to be met by the thermocouple is an unvarying signature that changes little with use (see section 4.3). The most difficult task during such a calibration may be the estimation of an uncertainty component to cover likely changes in use, if appropriate.

Before seeking or conducting a calibration there are other considerations that need to be addressed.

- The question of whether an *in-situ* or a laboratory calibration is preferable. In this context, a laboratory is any place having an environment specifically arranged for calibration. It therefore allows better access to the thermocouple tip and it offers reduced measurement errors and cleaner conditions for the temperature standard. It also transfers the responsibility of measurement to more specialised personnel. On the other hand, an *in-situ* calibration directly measures the interaction of the thermocouple's signature with the temperature profile and includes the effects of various systematic errors (section 5.3), such as those of the extension lead, the instrument and any stable DC contribution from AC noise.

- Used thermocouples require a different approach to new ones because their thermoelectric signatures are likely to have altered. They must be calibrated *in situ* or, if calibrated in a laboratory, have their signatures assessed first with a scan (see page 126). If a scan reveals significant irreversible change a laboratory calibration will usually be of little value and an *in situ* calibration is then preferable.

- Base-metal and rare-metal thermocouples require different treatments. They differ in accuracy level and in the extent to which they are affected by irreversible change. Thus, re-usable, variable-immersion standard thermocouples should be restricted to temperatures below about 500°C, for base-metal thermoelements, and to 1300°C, for rare metals. The limit for type N and type K thermocouples may be extended to at least 800°C if an ID-MIMS system is used (section 2.6.5). With proper care, the only changes that occur are reversible and the calibrated state can be recovered at any time by annealing or quenching (whichever applies: see section 4.3).

- Is the calibration to apply only to the moment of calibration, to a single brief use after calibration or to a longer period, say 100 h? Calibrations done *in situ* or for acceptance testing may not apply after the calibration because of drift in use.

4.3 Reference standard thermocouples

Thermocouples, as calibrated reference standards, are used for measurements taking up to 1 h at each of several temperatures, and often they are re-used in different furnaces and/or at different immersion depths. Values of temperature obtained with a reference standard are meaningful only if its Seebeck coefficient changes by an insignificant amount and its thermoelectric signature reflects an adequate homogeneity in coefficient.

As detailed in Chapter 2, the Seebeck coefficients of all thermocouples change at temperatures above $\sim 250°C$. They undergo thermoelectric hysteresis, on cycling the temperature, a consequence of reversible processes, and progressive irreversible change at higher temperatures. How then are thermocouples to be used as reference standards? The answer is to restrict the irreversible changes, especially by limiting the temperature range of use, and to control the reversible processes with either a quench or an anneal (see below). Let us consider the matter in three temperature regions, as follows.

Below 250°C Most thermocouple types, including K, N, S and T, have stable Seebeck coefficients and may be calibrated to an uncertainty better than 0.1°C + 0.1% of temperature over much of this range. At this accuracy level it is necessary to scan the thermocouple, at say 200°C, to estimate its inhomogeneity (as this sets the lower limit).

Below 500°C Here, reversible changes in Seebeck coefficient occur, peaking near 450°C. Thus thermocouples of type R (or S), N and K, initially in the quenched state, will drift approximately 0.1, −0.4 and 1.0°C, respectively, in about 100 h at a tip temperature of 450°C. On re-using them at a reduced immersion, so that emf is produced by wire previously at 450°C, the emf will change by the equivalent of 0.3, −2 and 4°C, respectively. On the other hand, the change in Seebeck coefficient may first be saturated by annealing the alloys at the temperature of maximum effect. For example, a uniform-temperature anneal for 16 h at 450°C would reduce the above changes by a factor of perhaps five.

Above 600°C At these temperatures any change that had occurred previously at temperatures around 450°C will reverse, and in types R, S and

4.3. REFERENCE STANDARDS

possibly N, but not in type K, a second reversible change will develop (sections 2.5.1 and 2.6.5). Also, irreversible changes in the coefficient will occur that limit the usefulness of the thermocouples as reference standards. Type K and type N thermocouples should be limited in use as reference standards to temperatures below about 500°C (800°C if ID-MIMS) and the bare-wire types B, R and S, to 1300°C. If a thermocouple's temperature is kept below these limits it is possible, with suitable anneals, to repeatedly return both its homogeneity and its emf-temperature relationship to that described by its calibration.

Reference standards used above the limits set by irreversible change, 500 and 1300°C (above), would be less accurate and should in general be restricted to one immersion depth.

4.3.1 Hysteresis state: quenched?

Let us examine the performance possibilities for bare-wire type R (or S) thermocouples, with temperatures kept below 1300°C. They will be considered, first in the quenched state and then in the '450 state' (described below). All comments made about type R thermocouples apply equally to the type S.

In this discussion, a **quenched state** refers to that condition of a thermocouple achieved by quenching it in air (near 20°C) from a temperature above the maximum temperature for hysteresis, viz, ~1050°C for types B, R, S and N and 700°C for type K. Most thermocouple wires and MIMS cables are quenched at the last stage in manufacture. Furthermore, it will be assumed that Pt-based reference standards are given a quench while in their twin-bore insulators. This is achieved by taking the ceramic-insulated length of thermocouple to a temperature of between 1050°C and 1100°C, and then quickly withdrawing it from the furnace, in a few seconds. It is an insulator-limited quench, less severe than if un-insulated and more typical of that experienced in use.

In a quenched state and fixed in position, the drift for a type R thermocouple is greatest above 1000°C, where it is ~ 0.1°C during the first 100 h. If kept below 1300°C drift is due to two reversible processes in different regions within the temperature gradient zone (see Figure 2.5 on page 37). There is low-temperature hysteresis in the section centred on 450°C and high-temperature hysteresis, centred on 800°C, and the effects are in opposite directions. Thus, partial compensation occurs. Larger errors of measurement may occur if the thermocouples are then moved or re-used. Consider thermocouples calibrated in the quenched state, used for some time, say 100 h, at the temperature of greatest reversible change (800°C, for type R) and then re-used at this temperature, or higher, at a reduced immersion depth. Then the error for a type R thermocouple could be 1°C.

These errors are minimised if the use of a reference standard is managed so as to avoid those combinations of temperature and immersion that result in the larger errors. For example, when a series of measurements are to be made with a type R thermocouple it is better not to finish at 800°C, if the next use requires a smaller immersion depth. Nor is it advisable for a reference standard, of any type, to be left in a furnace to cool—too much time may then be spent in the hysteresis zones. It would be better to raise the temperature of the type R thermocouple to at least 1000°C, increase the immersion, if possible, and leave it at temperature for 5 minutes or so before withdrawing it and have it cool rapidly to ambient. Better still, the thermocouple could be removed and given a recovery treatment in a suitable laboratory furnace, like that shown on page 128. It could be annealed for 5 to 10 minutes at 1050 – 1100°C, to reverse any residual hysteresis, and then quenched, to re-establish the conditions of calibration. Indeed, the strategic use of this anneal/quench procedure could repeatedly return the thermocouple to its calibrated state.

Alternatively, reference standards may be calibrated in the **450 state** [110]. To achieve this the type R thermocouple is placed in a uniformly quenched state, as described above, then uniformly heated overnight (16 h or so) at 450°C. It is important that during the latter anneal all that length of thermocouple to later experience temperatures above ambient, and thus likely to generate emf, be at 450°C. The initial *in-situ* drift for type R thermocouples in this state will be about 0.2°C in 100 h at 1000°C, marginally worse than for the quenched state (see above). The effect of re-using or moving the thermocouple after a substantial use is much the same as that mentioned for a quenched thermocouple.

Once again, the condition at calibration can be re-established at will, by the strategic use of annealing, and the errors minimised. To reproduce the 450 state, the active length of type R and S thermocouples could be held at 1050 to 1100°C for 5 to 10 min, to dissociate rhodium oxide (section 2.5.1), given a quench and then annealed at 450°C for 16-20 h.

From the above considerations it appears that Pt-based reference standards are better kept in the quenched state. Their drift is marginally less and the recovery procedure is easier. However, some national standards laboratories prefer to maintain their reference standards in the 450 state.

4.3.2 Calibration of the standard

Before calibrating a reference-standard thermocouple (type B, R or S) it should be placed in its most thermoelectrically homogeneous and stable state. This

4.3. REFERENCE STANDARDS

requires four steps.
(1) a cleaning procedure (see section 4.3.3),
(2) high-temperature annealing to thermoelectrically stabilise the thermoelements,
(3) annealing for $\sim$ 1 h at 1100°C to remove the effects of handling (cold work), once the thermocouple is assembled in its alumina insulator, and
(4) quenching, or annealing to the 450 state (see above).

The high-temperature stabilisation of type S, and thus of type R, thermocouples was extensively studied at the National Research Council (NRC), Canada, by McLaren and Murdock [35, 111]. As a consequence, NRC have adopted 10 h at 1300°C as their preferred anneal—given to the bare wires by electric self-heating. On the other hand, the National Institute of Standards and Technology, USA (NIST) [112] and NML [113] apply 1 h bare-wire anneals at 1450°C to their best standards, with a further 1 h at 1100°C given by NML. The NML anneals increase the Seebeck emf at 400°C by 5 μV, equivalent to 0.13% of the temperature [37]. Other national standards laboratories have chosen different procedures, some no higher than 1200°C [112].

For its routine calibration service (section 4.3.3), NML has adopted a slightly less severe set of high-temperature anneals than those given above, and a compromise procedure is given below for those laboratories not wishing to apply bare-wire anneals.

Reference standards should be calibrated by the most accurate means available, i.e., with a calibration uncertainty dictated mainly by the inherent level of inhomogeneity in Seebeck coefficient along the thermocouple. The most accurate method is the use of suitable metal fixed-points and intercomparison with a standard platinum resistance thermometer, calibrated directly on ITS-90. Then, the 'expanded uncertainty' (defined on page 163) of calibration for types R and S, expressed as 95% confidence limits, would be typically 0.3 μV + 0.025% of emf [114], equivalent to 0.22°C at 1000°C for type R. This value includes 0.02% as the inhomogeneity component and applies to temperatures of use no greater than 1100°C.

It is far simpler, and thus less expensive, to calibrate by intercomparison with a thermocouple previously calibrated as above and carefully maintained in its optimum condition, such as the approach taken by NML (see below). The resultant calibration uncertainty is larger, about 0.9 μV + 0.035% of emf, and the method allows extension to higher temperatures. In most cases the increase in uncertainty is of little concern.

Pt-based thermocouples intended **for less-accurate work** may be cleaned and annealed in a less rigorous manner. To clean the wires a nitric acid wash followed by an alcohol rinse may be used or a wipe with acetone then alcohol. Then, rather than apply a high-temperature, bare-wire annealing procedure (needing a specialised facility), the insulated anneal (step 3 above) could be extended as a compromise. I suggest the cleaned wires be assembled in a new length of twin-bore alumina, annealed for 2 h at 1150°C and quenched from this temperature. It puts the thermocouple in a 'known', stabilised state, maybe 0.05 to 0.1% higher in emf than the 'as-new' state, provided it is not then used above 1100°C. Use above this temperature could cause a further change of about $2\,\mu$V at 232°C, equivalent to 0.1% of temperature [35]. The need for further annealing depends on the state required for calibration, and finally, the thermocouple should be calibrated by a method such as that outlined in section 4.5 using a reference standard with a suitably small calibration uncertainty.

4.3.3 NML calibration procedure

Types B, R and S thermocouples, in 0.5 mm diameter bare-wire form, are calibrated at the CSIRO National Measurement Laboratory in the quenched state. The NML calibration procedure is fully described elsewhere [115]. Briefly, it is as follows.

Each thermocouple is dismantled and the wires are cleaned by placing them for 15 min in each of the following boiling liquids: nitric acid (20% of full strength), hydrochloric acid (20%) and distilled water. After each acid wash the wires are rinsed in distilled water. They are bare-wire annealed (electric self-heating) at 1400°C and then at 1100°C, in each case for 30 min. The electric (heating) current is then linearly reduced to zero over 1 min, to approximate the quench given later when insulated. After installing in a single length of new, twin-bore, re-crystallised alumina the thermocouple is thermoelectrically scanned (see page 126), if previously used, and its ceramic-insulated length is given a uniform anneal at about 1100°C for 10 min. Then, it is quickly withdrawn from the furnace to achieve the quenched state defined on page 119. The wires are handled as little as possible and then only with clean tweezers and disposable plastic gloves and on a clean glass-topped bench.

The calibration to 1100°C is automatic and quick (10°C/min on heating)— it involves intercomparison measurements with an NML standard thermocouple, previously calibrated at the fixed points [113], and a type B check thermocouple. The latter thermocouple serves as a run-to-run continuity check and as a sensitive guide to any change in DVM calibration [114]. For higher temperatures the calibration is extended by measurements at the melting point of palladium in argon (1554.8°C [23]) and, if necessary, at the melting point of platinum (1768.2°C).

For a temperature range of 0 to 1100°C, or 0 to 1550°C, the calibration of new wire would have an 'expanded uncertainty' (see page 163), expressed at a 95% level of confidence, of $(0.9\,\mu V + 3.5 \times 10^{-4} V)$, for a thermocouple output of V (μV), e.g., $4.6\,\mu$V at 1000°C for type R, equivalent to 0.35°C. The corresponding uncertainty for a calibration limit of 1750°C is $(0.9\,\mu V + 3.5 \times 10^{-4} V + 0.85 \times 10^{-8} V^2)$. These values apply to wire having an as-new level of inhomogeneity of ±0.02% of emf—for wires with greater inhomogeneity the calibration uncertainty would be larger.

It should be noted that the reported calibration uncertainty covers the likely errors of measurement during the calibration and the effect of using a different immersion depth (within a defined length, the 'calibrated length'). However, it does not cover the changes that may be expected in any subsequent use. Over a 100 h period of use at temperatures up to 1100°C, in differing immersions depths and/or in different furnaces and removed from furnaces while at different temperatures, the calibration is likely to change, up and down, by about $\pm(4.5 \times 10^{-4} V)$. This is an estimate to 95% confidence and applies only to clean oxidising conditions. It is also a reasonable estimate for temperatures of use up to maybe 1300°C. In practice there are three approaches that may be taken for dealing with such change.

1. Simply include the above use-component when calculating the uncertainty of measurement, as done in the table on page 141.

2. Monitor the *in-situ* drift, for a single-immersion application and correct for the drift.

3. Subject the thermocouple to a suitable annealing/quench sequence, as required, to keep it in its as-calibrated state (see above). In this case, long term effects should be checked by scanning the thermocouple at regular intervals or whenever in doubt.

4.4 Laboratory calibration

The meaning adopted for the term 'laboratory calibration' is given in section 4.2.1, beginning on page 115, which also deals with calibration issues, such as, the need for scanning and whether an *in-situ* calibration is more appropriate. The possible need for pre-annealing or quenching should also be considered (see pages 118 to 120).

During a calibration the tip temperature must be either established in a temperature-defining fixed-point, usually a metal freezing point, or measured by another (calibrated) sensor. The expense and complication of using a series of fixed points is warranted only in special circumstances. Consequently, the following discussion will be confined to calibrations conducted as intercomparisons with a higher-level reference standard.

Sometimes it is desirable to use a Pt-resistance thermometer [116] as the temperature-reference standard. This is so, for example, if the required uncertainty of calibration is less than 0.5°C and the temperature limit for the calibration is less than about 250°C. Then, the calibration is best done in a stirred liquid bath [117] for temperatures above about −80°C, and in a cooled copper block, if lower. The procedure would be, firstly, to calibrate the thermocouple as a function of temperature in roughly 50°C increments over the desired range, at any convenient depth of immersion. Secondly, to scan the thermocouple if it had had prior use or when the required uncertainty is less than 0.3°C. Scanning is most easily done in a bath at the calibration temperature most removed from ambient (see page 126) or by progressively immersing it in a deep Dewar of liquid nitrogen, if appropriate.

On the other hand, most thermocouples should be calibrated by comparison with a thermocouple reference standard, especially if the upper temperature of use and calibration exceeds 300°C. It is the most practical form of calibration and, in its most accurate form, it is the simplest. Both the standard and the unknown, welded or wire-wrapped together at their tips, are run unbroken to an ice pot (at 0°C) and from there the circuit is continued in copper to a switch and a digital voltmeter (DVM). The calibration uncertainty for such an arrangement may be judged from the example beginning on page 140.

The arrangement may be made more convenient by installing in-built extension wires (not compensating wires—page 85) for the standard and for the various types of thermocouple to be calibrated, and using the switching circuit of Figure 3.5. The thermocouples can then be shorter and the extra uncertainty introduced by the extension wires is small—indeed, it is insignificant if sufficient care is taken. Moreover, a calibration using extension wires is recommended for thermocouples normally used with their CJ ends at ambient temperature. If such a thermocouple is calibrated with the ends in an ice pot (0°C) significant emf will be generated at the entrance to the ice pot, in wire that normally does not contribute to the measured signal—wire that is most likely cold-worked and possibly not given the same annealing treatment as the remainder of the thermocouple, being at the end. In this case, emf produced in the ice-pot region would contribute error unnecessarily to the calibration.

The effect of extension and compensating wires on calibration is given in Table 4.1, for an assumed set of conditions. It is given in the form ($X\,\mu V + Y\%$). The first component, X, refers to the 0°C-to-ambient section of the lead and is independent of tip temperature. The second, Y, given as a percentage of the emf equivalent of the tip temperature, is calculated by assuming the temperature of the 'head', where the thermocouple connects to the lead, rises

4.4. LABORATORY CALIBRATION

Table 4.1: The uncertainty (see note on page 172) introduced by using extension or compensating wires in calibration, estimated at the 95% level. The values apply only to the assumed conditions—that the head temperature increases with tip temperature by 1 in 50 (see text).

Thermocouple Type	Extension Wires	Compensating Wires
B	1 μV	3 μV*
S or R	1 μV + 0.01%	15 μV + 0.1%
N or K	15 μV + 0.03%†	not appl.

† assumes the use of premium-grade thermocouple wire for extension lead.
* for Cu versus Cu and head temperature < 50°C.

by 1°C for every 50°C the tip temperature exceeds that of the ambient. The assumption is merely a convenient means of estimating the effect of a changing head temperature and ought to be an over-estimate, since it is unwise during calibration to allow the head temperature to vary to this extent. It is easy to keep the head at ambient temperature and so avoid component Y.

Furthermore, it is assumed that premium-grade thermocouple wires are used as the extension lead, preventing the calibration from being compromised by a low-cost component of the system. The uncertainty for extension wires may be improved further by calibrating the wires and selecting those with insignificant corrections up to say 60°C.

I strongly recommend that thermocouples be calibrated with their tips in intimate contact with the tip of the standard—preferably welded, but otherwise wire-wrapped together. A suitable wrapping technique is described in reference [118] and illustrated in Figure 4.1. For the three MIMS probes shown the temperature difference between thermocouple tips is less than 0.1°C at 500°C, i.e. < 0.02%, when the tips are in a region uniform to $< 0.5°\text{C cm}^{-1}$. These results would apply to the furnace of Figure 4.2.

The Pt-based standard will be contaminated by such calibrations. Nevertheless, if the heated section of the standard is within a continuous length of twin-bore alumina the wires will be contaminated only where they contact the Nichrome wrap and other thermocouples—near their tip. Here, if we assume the Seebeck coefficient of the standard is changed by 10%, over a length of 5 mm, contamination will contribute an error of $\sim 0.03°\text{C}$ to any measurement in a temperature gradient of $0.5°\text{C cm}^{-1}$. This error is not significant and in any case the repeated use of a wrap eventually breaks the Pt-based wires and the contaminated section must then be cut away to re-weld the tip.

Periodically the standard should be checked to see if any significant change has occurred. For this check the tip and $\sim 2\,\text{mm}$ of adjacent wire should be

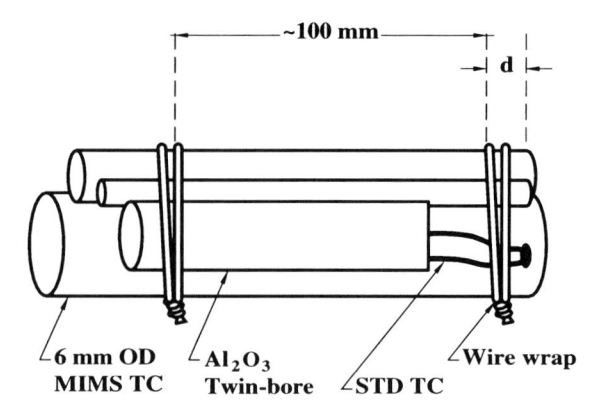

Figure 4.1: A schematic showing the 'wire-wrap' method of binding a standard thermocouple (STD TC) to three MIMS probes, of 1.5, 3 and 6 mm diameter. Dimension d should be 2 to 3 mm—see ref. [118].

removed and the standard placed in a length of new twin-bore alumina and scanned at 450 °C.

Sometimes a laboratory calibration is done with the standard thermocouple isolated from the others to avoid the risk of contamination or because the design of furnace prevents the thermocouples being together. Then, the thermocouple tips are separated and their temperatures will differ. When in a temperature-equalising block, the difference will contribute to the calibration an error of about ±0.1% of the tip temperature (unless within a heat pipe [117]). Moreover, the temperature uniformity of the block needs periodic re-assessment to determine the likely errors—not an easy exercise, especially if the immersion of test probes is changed to measure axial variations. Is the measured change, on varying the immersion, due to a temperature difference or to inhomogeneity? At each re-assessment there needs to be dummy wires or thermocouples present to serve as typical heat loads.

4.4.1 Thermoelectric scanning

While the level of inhomogeneity of new thermocouple wires may be safely assumed, that of a used one cannot, and the thermoelectric signature of a used thermocouple must be determined with a scan before calibration. This will establish whether its use has significantly affected the Seebeck coefficient and, if changes are evident, the signature will give a direct measure of the inhomogeneity component needed when calculating the uncertainty of calibration. Alternatively, it may suggest that the thermocouple be rejected.

If the thermocouple under consideration is restricted in use to temperatures

4.4. LABORATORY CALIBRATION

below about 250°C the calibration is best done against a Pt resistance thermometer in a stirred bath, as discussed above (page 123). At such temperatures, and especially if kept below 200°C, the Seebeck coefficient is relatively stable and an as-new, base-metal thermocouple may be calibrated and used to better than 0.1°C—the uncertainty achieved being dictated ultimately by its level of inhomogeneity. For an uncertainty of 0.3°C, or better, an assessment of inhomogeneity is needed, even for new wires. It is also necessary if the thermocouple had received any use that could have caused a change in its Seebeck coefficient, e.g., taken above 250°C, potentially contaminated or strained. For thermocouples restricted to temperatures below 250°C a scan is best done in a stirred liquid bath at the calibration temperature most removed from ambient, so developing the maximum thermocouple emf, and thus the extent of its fluctuation (inhomogeneity). The scan is done by calibrating the thermocouple as a function of immersion in 20 to 50 mm increments over all the length likely to be 'active' in a future application. Obviously, the bath temperature should be held constant during the scan. The thermocouple may be scanned while in a glass sheath, taking care to avoid significant 'conduction error' (at short immersion, see page 177) and to ensure that the bath is sufficiently uniform in temperature. Alternatively, a flexible thermocouple may be placed in direct contact with the bath liquid (alcohol down to -70°C, water up to ~ 95°C, if in impervious insulation, or silicone oil, to 250°C, with some cosmetic damage). Then, by keeping the thermocouple tip fixed in position, preferably wire-wrapped to the resistance thermometer, the immersion may be increased arbitrarily by progressively inserting (and flexing) more thermocouple into the liquid.

For thermocouples used above 250°C thermoelectric scanning may be achieved by comparison with another thermocouple in a furnace. A good general-purpose scanning facility, also suitable for thermocouple calibration, is described elsewhere [118]. It is based on a long tubular furnace (Figure 4.2), uniform in temperature to $\pm 1\%$ over much of its length, and a homogeneous reference thermocouple. The tip of the thermocouple under test is welded or wire-wrapped to that of the Pt-based reference thermocouple, referred to below as the 'scan standard'. The latter does not need to be calibrated, although it does need to be cleaned and uniformly annealed (section 4.3), assembled in a single, long (≥ 1000 mm) twin-bore insulator and stabilised by annealing its insulated length for at least 24 h at 450°C. The scan standard should not be used for any other purpose and not be taken above 450°C. The thermocouple is scanned by calibrating it against the scan standard as a function of immersion depth in steps of less than 50 mm. After each change in position sufficient time should be allowed for the temperature profile to be re-established along the wires (the computer-controlled facility at NML increments the thermocouple by 4 mm every 80 sec and collects a data set at the end of each pause).

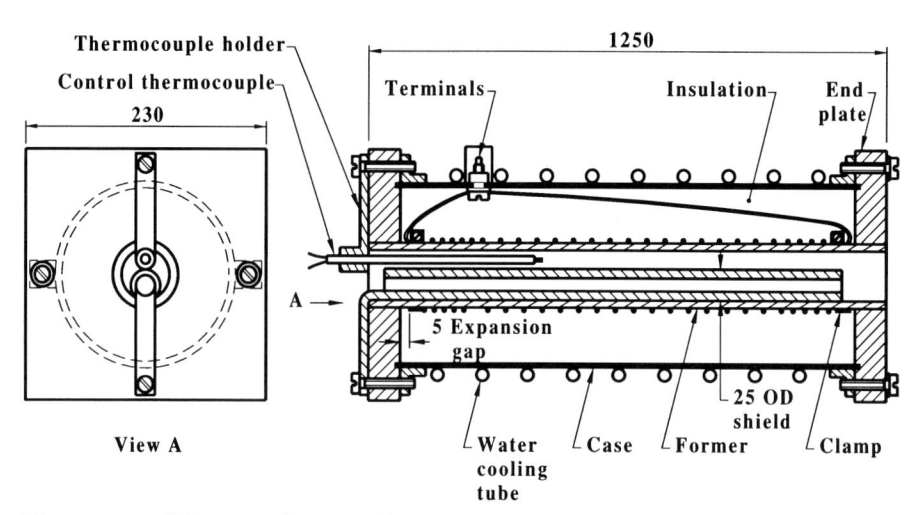

Figure 4.2: Schematic diagram of a wire-wound furnace suitable for scanning and calibration, with dimensions given in mm—from ref. [118]. In use, thermocouples are immersed from the right hand end into the earthed Inconel shield.

I suggest the result of a scan at 450°C be a plot of $(V - V_{ref})\frac{450}{T}$ against depth of immersion, where V_{ref} is obtained from the temperature of the scan standard and T is the temperature at each measurement of thermocouple emf V. The factor $450/T$ is a barely-significant correction to approximately compensate for changes in temperature of up to 10%, as the throat-gradient at either end of the furnace is approached.

The temperature of the scan furnace is important: 450°C is recommended. If the thermocouple is in its 450 state (annealed at 450°C, see page 120) it would be unaffected by the scan, and any part or all of the scan can be repeated at will. Moreover, the scan standard will remain homogeneous and will not require a 'recovery' anneal after each use in the scan furnace. Alternatively, if the thermocouple under test is in its quenched state (page 119) the scan could still be done—the result will correctly indicate inhomogeneity, if immersed at a steady rate, but its signature would progressively alter. It is preferable to first anneal the thermocouple at 450°C, then scan it and finally re-establish its desired state by a re-quench.

The scans for three type R thermocouples are shown in Figure 4.3. To convert the observed range in emf to an uncertainty component in the calibration it should be halved and expressed as a percentage of the emf generated by the thermocouple in the temperature gradient zone of the scan furnace, i.e., from the furnace temperature to ambient. The uncertainty components for thermocouples 1 and 2 of the figure are 0.015% and 0.08%, respectively.

Thermocouple 3 is considered unsuitable for re-calibration.

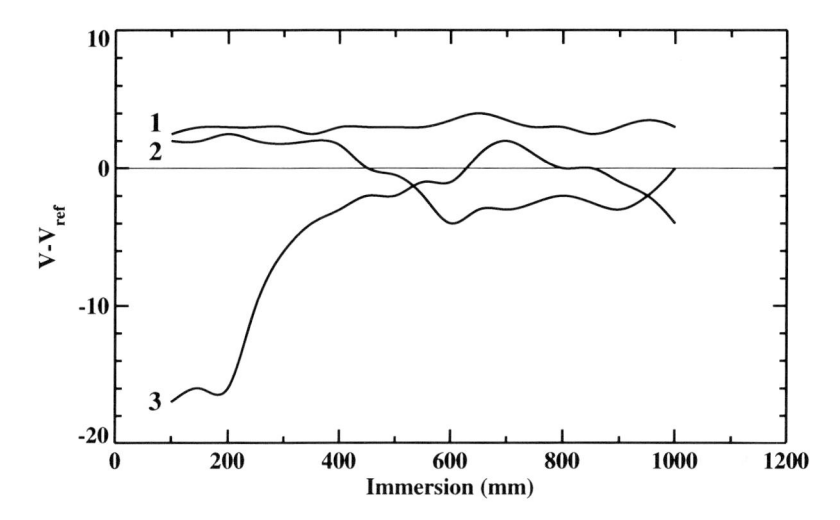

Figure 4.3: Thermoelectric signatures of three type R thermocouples—the change in $V - V_{ref}$ (μV) as a function of immersion depth at 450°C. Thermocouple **1** was new, thermocouple **2** had been used as a standard up to 1100°C for 100 h and thermocouple **3** had some use near 1500°C.

A scanning facility is also a useful diagnostic tool. Thermocouples that seem to produce odd signals in service could be brought to the laboratory and scanned at a suitable temperature, and the signature analysis will show whether the thermocouple has any irregularities. Notice, however, that the reference thermocouple for diagnostic tests should not be the one used for scanning in the calibration process (the scan standard), because of the risk of contamination and the potentially different scan temperatures.

4.5 A suggested calibration procedure

A common reason for the calibration of thermocouples is to check whether a particular specimen, or the batch from which it came, conforms to the manufacturer's tolerance on thermocouple output, normally expressed in temperature units. The broadest tolerance for base-metal alloys is ± 2.2°C or ± 0.75% of temperature, whichever is the greater (see table on page 28). To check a thermocouple, even to this level, is not a trivial exercise (see below).

In using calibration data to decide whether a thermocouple complies with the specified tolerance, it would be unfortunate if the decision to 'fail' or 'pass' the thermocouple was to a large extent affected by the calibration uncertainty itself. It is important then for the uncertainty of calibration to be relatively small, say, no larger than one fifth of the tolerance with which the thermocouple is being compared. Thus, for the above standard-grade tolerance, an upper limit for the calibration uncertainty would be about 0.44°C up to 290°C and 0.15% thereafter.

A relatively straightforward calibration facility would be one in which the thermocouple under test is directly compared with a Pt-based standard thermocouple in a furnace capable of operating over any desired range. With care, an uncertainty of $(0.15°C + 0.13 \times 10^{-2}T)$ may then be achieved in the calibration of a base-metal thermocouple (see Table 4.5). This level of uncertainty just meets the above-stated needs of a compliance test for standard-grade tolerances. Moreover, the uncertainty is little more than the contribution from inhomogeneity along the thermocouple under test (0.1%, assumed). Obviously, at this level, the calibration procedure must contribute little else to the overall uncertainty.

From the above considerations I suggest the following general-purpose procedure for thermocouples (including the rare-metal types)—to be read in conjunction with section 4.4. It is suitable for compliance testing and for calibrations at any set of temperatures up to at least 1100°C. It should **not** be used to calibrate a **used** thermocouple unless the latter has been thermoelectrically scanned (page 126) and shown to have an insignificant level of inhomogeneity. Of course, if a thermocouple is scanned before calibration the variation so measured should be included as the inhomogeneity contribution (half the measured range, expressed as a percentage of temperature) to its calibration uncertainty. I suggest:

- The new, or as-new (see above), thermocouple to be calibrated and a standard thermocouple are inter-compared in a tube furnace of the type shown in Figure 4.2 (page 128).

- The standard should be Pt-based and calibrated over an appropriate temperature range to the best available uncertainty. Its calibration uncertainty should apply to any immersion depth (i.e., the calibration laboratory included a component for inhomogeneity) and any drift (through use above about 500°C) since its most recent calibration must be covered by an appropriate component of uncertainty (see page 123).

- The tips of the standard and the test thermocouples should be welded or wire-wrapped together (section 4.4) to avoid a significant temperature difference. This also overcomes the need to test the calibration furnace for spatial-uniformity in temperature.

- Compensating wires should not be used—extension wires (see page 85) may be, if tested and shown to contribute insignificant error over the

4.5. A SUGGESTED CALIBRATION PROCEDURE

temperature range of its use (I suggest the temperatures of head, where the thermocouple and lead join, and CJ be kept within about 5°C of each other).

- A Digital Voltmeter (DVM), or a voltage range on a DMM, should be used to measure the emf's, rather than an instrument that displays in temperature units. The former costs significantly less and is easier (cheaper and quicker) to calibrate. Since a DVM does not need to measure and correct for the CJ temperature of each thermocouple and doesn't convert emf to temperature, its calibration uncertainty is less.

- Compensation for CJ temperatures and switching should be done using a method similar to that depicted in Figure 3.5.

- The conversion of emf data to values of temperature is best done by software using the standard reference equations given in Appendix B (see comments on page 220). Then, the conversions to temperature may be done without contributing to the calibration uncertainty. Alternatively, the conversion should be done manually.

- The outcome of the calibration—results suitable for a report—may be generated using the interpolation procedures outlined in section 4.6.

- Calibration data should be accompanied by an expression of uncertainty—in Australia, evaluated at the 95% probability level (see section 5.4).

- When calculating the calibration uncertainty (as a guide, see Table 4.5 on page 141) it will usually be sufficient to presume that the contribution from inhomogeneity in as-new wires is 0.1%, for base-metal thermocouples, and 0.02%, for rare-metal. On the other hand, if a base-metal thermocouple is being compared with special limits of error ('premium-grade' wire with a tolerance half that of standard-grade wire) the calibration uncertainty may then be considered excessive (roughly half that of the special-grade tolerance). If so, and since the main contributor to the calibration uncertainty is the presumed level of inhomogeneity, the thermocouple should be scanned. In many cases, the measured level of inhomogeneity will be less than indicated above. Then the calibration uncertainty will be reduced, and it should be noted that it then relates **only** to that thermocouple—it does not indicate the possible variations along the production coil it may represent.

- It follows from the above that if there is a reasonable chance that thermocouples will need to be scanned the calibration facility ought to include such a feature. For example, the calibration furnace should have an appropriate length and a special homogeneous, reference thermocouple is required (see page 126). Of course, once the ability to scan thermocouples is included the facility becomes a useful diagnostic tool.

If the thermocouple is one of a batch with a common purpose, such as that required for enclosure testing (Chapter 6), see the section "Movable test probes" on page 143.

4.6 Interpolation

During a laboratory calibration each thermocouple under test has its CJ either at 0°C or corrected to 0°C. Its tip temperature is accurately measured and the calibration data consist of paired values of emf and temperature, obtained at several, somewhat arbitrary, values of temperature covering the range of interest. To fully characterise the behaviour of the thermocouple, values of emf at other temperatures need to be calculated by interpolation. Consider the following alternatives for interpolation.

- The most difficult and least reliable method is to plot the raw data: emf versus temperature. For a type R thermocouple calibrated to 1000°C, for example, and with a resolution of $1\,\mu$V (1 mm of graph), the graph would need to be more than 10 m long to contain all the data!

- The simplest method is to manually produce a 'difference curve' (plot the deviation function). Most of the functional character in the relationship between emf and temperature for the thermocouple will be contained in the relevant standard reference equations. Its emf for a CJ temperature of 0°C, V, will differ only slightly from the corresponding reference tables value V_{ref} and so the difference $(V - V_{ref})$ will be small and easy to plot. Also, the shape of the curve of best fit should be a simple, slowly changing function of the temperature, T.

- A computer method: there are many methods available and probably the most reliable are those that fit $(V - V_{ref})$ to T, or to V, for the reasons given above—an example begins on page 135.

As an example of the manual approach, consider the calibration of a type K thermocouple against a type R standard over the range 0 to 1000°C. Assume that at various test temperatures, chosen to adequately cover the range of interest, 'simultaneous' measurements of V and V^o (the standard) are made. The main requirement in selecting test temperatures for calibration is that they be sufficient in number and so spaced that the temperature dependence of $(V - V_{ref})$ can be reliably determined. There is no advantage in trying for

4.6. INTERPOLATION

Table 4.2: Sample data for the calibration of a type K thermocouple using a type R standard and manipulated in emf units. The same procedure could have been followed using temperature units (see Table 4.4).

Measurements*		Calculations				Results
Thermocouple	Standard					
V	V^o	$(V^o - V^o_{ref})$	V^o_{ref}	T	V_{ref}	$(V - V_{ref})$
(μV)	(μV)	(μV)	(μV)	(°C)	(μV)	(μV)
2 032	291	0	291	49.1	1 986	46
4 538	716	−1	717	109.2	4 476	62
7 672	1 353	−1	1 354	186.9	7 615	57
13 043	2 580	0	2 580	318.3	12 969	74
19 123	4 064	1	4 063	462.2	19 035	88
26 668	6 016	2	6 014	637.6	26 501	167
34 880	8 372	4	8 368	833.7	34 652	228
41 850	10 617	7	10 610	1007.9	41 583	267

* calibration corrections for the measuring instrument, e.g., DVM, have already been applied.

temperatures at the cardinal points, 100, 200,...,1000. Indeed, it would be a waste of effort.

A possible measurement set for the calibration is shown in Table 4.2. What is needed from the calibration is the difference between the signal generated by the thermocouple, V, and the emf, V_{ref}, given in the type K reference tables for the same temperature. This is done in two stages: the tip temperature T is calculated from the reference thermocouple emf, V^o, and V_{ref} is obtained from T. To get T the calibration error of the standard, $(V^o - V^o_{ref})$, is interpolated from its calibration report and the reference tables emf, V^o_{ref}, for the same, as yet unknown, temperature is obtained by the obvious arithmetic: $V^o_{ref} = V^o - (V^o - V^o_{ref})$, that is, column 2 minus column 3. Columns 5 and 6 also require the reference tables: those for type R thermocouples to get T from V^o_{ref} and those for type K to convert T into V_{ref}. Finally, $(V - V_{ref})$ is simply the difference between columns 1 and 6.

The data are then plotted, as in Figure 4.4, and a smooth best-fit curve is drawn. The aim should be to produce the simplest curve possible, consistent with the data—scatter should reflect the experimental set-up and faulty data will be highlighted. If in doubt calculations should be checked and, if necessary, some measurements repeated.

The curve more accurately represents the behaviour of the thermocouple than do the raw data. Therefore, for the calibration report we supply curve values of $(V - V_{ref})$ at selected temperatures. The report should also contain the values of emf it would generate at these temperatures, calculated using the layout in Table 4.3. Here, the curve values of $(V - V_{ref})$ appear in column

2 and the values of V_{ref} in column 3 were obtained from the type K reference tables at temperatures T. Finally, V is simply the sum of columns 2 and 3.

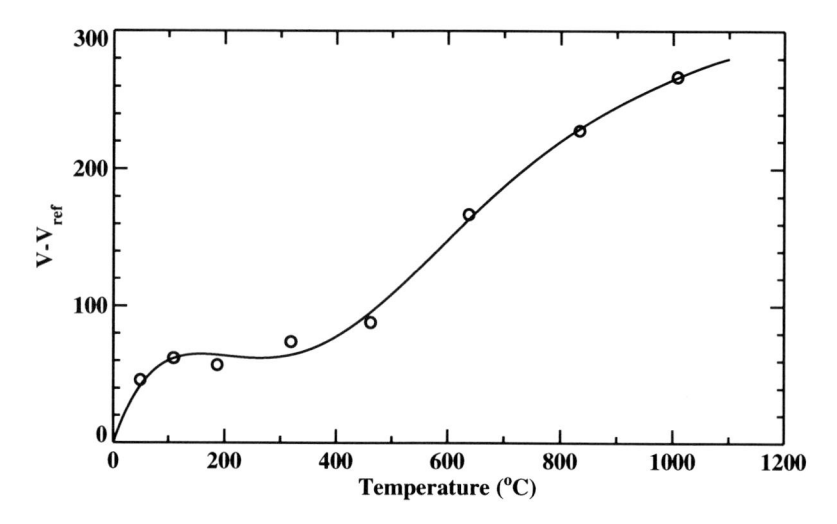

Figure 4.4: Plot of calibration data, $V - V_{ref}$ (μV), from Table 4.2, and a best-fit curve (polynomial).

In selecting the temperature interval ΔT for reporting calibration data (e.g., 100°C for column 1 in Table 4.3), it should be remembered that the user will probably use linear interpolation to get $(V - V_{ref})$ at intermediate values of V or T. Consequently, ΔT must be chosen for insignificant interpolation errors. Consider the possibility of using a reporting interval of $\Delta T = 200$°C. In Figure 4.4 the most non-linear, 200°C-wide region is from 0 to 200°C. Here, a straight line from the origin to 64 μV at 200°C would pass through 32 μV at 100°C, yet the best-fit curve yields 52 μV at this temperature. Thus, if data had been given in intervals of 200°C a user could linearly interpolate at 100°C and get 32 μV for $(V - V_{ref})$, in error by -20 μV.

The final report would not contain column 3 of Table 4.3 and the emf values need to be rounded to a level appropriate to the uncertainty of calibration (section 4.7). A recommended rounding procedure is described in section 5.5.

Calibration data were treated above in emf units (Table 4.2) leading naturally to values of calibration error, $(V - V_{ref})$, in μV. This was done because the signal produced by a thermocouple is an emf and it is logical and consistent to conduct the calibration in terms of emf. However, a thermocouple is also a 'temperature sensor', and presumed to be used in conjunction with the reference tables. Thus, it is equally valid to conduct the calibration in temperature units. It involves just as many operations, and may be less

4.6. INTERPOLATION

Table 4.3: Values of thermocouple emf, V, and its calibration error at selected temperatures, from Figure 4.4. Column 3 (not required in a calibration report) is included to show the intermediate step in calculating V. I suggest that V be included in a report to reduce the risk of misinterpreting the sign of the 'error'.

Temperature T (°C)	Thermocouple Error $(V - V_{ref})$ (μV)	Type K Tables V_{ref} (μV)	Thermocouple Emf V (μV)
0	0	0	0
100	60	4 096	4 156
200	64	8 138	8 202
300	63	12 209	12 272
-	-	-	-
-	-	-	-
1000	265	41 276	41 541

confusing—see Table 4.4.

In the table, the calibration error of the standard, $(V^o - V^o_{ref})$, assumed to have been expressed in μV in its calibration report, was converted to the equivalent in °C via columns 3 to 5. In practice, columns 3 and 4 are best avoided by converting any such calibration data for a standard into temperature units prior to use. Then, the conversion (i.e., $(T^o - T) = (V^o - V^o_{ref})/S^o$) would be done once only, calculated at each value of temperature for which the calibration error is reported, and suitable values of $(T^o - T)$ read or interpolated from this revised table when needed. Of course, if the calibration errors had been reported in temperature units the conversion, and columns 3 and 4, would not have been required.

The 'true' temperature T is the temperature measured by the standard, after correction, and is thus determined by subtracting data in column 5 from the corresponding values in column 2. Then, the error in the thermocouple being calibrated, $T' - T$ of column 7, is simply column 1 minus column 6.

4.6.1 Software for generating calibration results

If computing facilities are available calibration results may be generated with ease. The first step in preparing suitable software is to fit an interpolating equation to the calibration data and, as explained above, this is best done by representing the temperature dependence of the (small) difference, $V - V_{ref}$. One approach for getting a 'best-fit' is to use a third- or fourth-order, least-squares polynomial. Then, this equation, like the hand-drawn curve mentioned

Table 4.4: Sample data for the calibration of a type K thermocouple, using a type R standard, in temperature units (upper table)—directly equivalent to the data in Table 4.2 (using emf units). Here, S^o is the Seebeck coefficient of the standard. The results of calibration, suitable for a report, are taken from a best-fit curve equivalent to that in Figure 4.4 and rounded (lower table).

Measurements*		Calculations				Results
Thermocouple T' (°C)	Standard T^o (°C)	$(V^o - V^o_{ref})$ (μV)	S^o (μV/°C)	$T^o - T$ (°C)	T (°C)	$(T' - T)$ (°C)
50.22	49.16	0	6.5	0.00	49.16	1.06
110.70	109.08	-1	7.6	-0.13	109.21	1.49
-	-	-	-	-	-	-
-	-	-	-	-	-	-
1014.76	1008.38	7	13.3	0.53	1007.85	6.91

* calibration corrections for the measuring instrument have already been applied.

Temperature T (°C)	Thermocouple Error $T' - T$ (°C)
0	0
100	1
200	2
-	-
-	-
1000	7

above, would more accurately describe the thermocouple over the temperature range of calibration than do the raw data. Accordingly, it is used to generate the values given in the calibration report.

Care must be taken to ensure there is sufficient redundancy in the data to avoid fitting-artifacts, e.g., oscillations or excessive bias from faulty data. For a fourth-order fit to $V - V_{ref}$ I suggest that at least eight data points would be required. Moreover, the quality of fit should be assessed from fitting parameters, such as the rms deviate, calculated by the fitting routine.

In a report, values of the difference $V - V_{ref}$, and possibly the thermocouple emf V, are given as a function of temperature T at suitable increments over the desired range. The incrementing step may be large, e.g., 100 or 200°C (see above), if values of $V - V_{ref}$ are given, or small, say 1°C, otherwise.

For a measurement set consisting of n values of thermocouple emf, $V(i)$ ($i = 1, 2 \cdots n$), and the corresponding temperatures, $T(i)$, the program could begin by converting values of $V(i)$ into the equivalent differences, $D(i) =$

4.7. CALIBRATION UNCERTAINTY

$V(i) - V_{ref}$, where V_{ref} is that value corresponding to $T(i)$. Then, the data pairs, $D(i)$ and $T(i)$, are fitted using one of the readily available routines for generating a least-squares polynomial in the form: $D = \sum_{j=0}^{4} d_j T^j$. Finally, for the calibration report, values of thermocouple emf, V, are calculated every 1°C from 0 to 1000°C, as a convenient example, as follows.

Step			
1	:	for $i = 1$ to n	
2	:	$T = T(i)$	
3	:	GOSUB getVref	! sub-routine on p. 220: gets V_{ref} from T
4	:	$D(i) = V(i) - V_{ref}$	
5	:	next i	
6	:	GOSUB lsq-poly	! gets coeff's d_j—least-squares fit of data
7	:	for $T = 0$ to 1000 step 1	
8	:	$E = 0$ and $j = 4$	
9	:	$E \Leftarrow d_j + E$	
10	:	if $j = 0$ go to step 13	
11	:	$E \Leftarrow ET$ and $j \Leftarrow j - 1$	
12	:	go to step 9	
13	:	$D = E$	! the fit value $(V - V_{ref})$ at T
14	:	GOSUB getVref	! comment as per step 3
15	:	$V = D + V_{ref}$	
16	:	print T, V and/or D	! $D = V - V_{ref}$
17	:	next T	

Whereas the power of modern computers may be used to great advantage in measurement and calibration, for example, there are pitfalls. One is mentioned in item 17 on page 183.

4.7 Calibration uncertainty

The uncertainty of calibration for a thermocouple should include the following components, which fall logically into four groups. Each should be estimated and combined in the manner described in section 5.4 (page 159).

1. The uncertainty in the knowledge of the chosen temperatures. If using fixed points (e.g., metal freezing points) the uncertainty represents how well each temperature is realised, i.e., the effect of purity, the dynamics of the freeze, etc. On the other hand, for a calibrated temperature sensor, the reference standard, we have:

    (a) The calibration uncertainty of the standard.

    (b) The uncertainty in the use of the standard, including its potential drift since calibration.

Note: in general a calibration uncertainty (for thermocouple, instrument, etc) is not to be taken as the uncertainty in use, subsequent to calibration, unless otherwise stated in the calibration report. Usually, environmental and other conditions are optimised during calibration to minimise the uncertainty of calibration, conditions that may not apply in use. For example, a temperature-indicating instrument is often calibrated with its ACJC circuit bypassed ("External ACJC" chosen).

(c) The uncertainty in the measurement of the standard. The measuring instrument, assumed here to be a digital multimeter (DMM) set on a suitable mV range, will contribute the following components (see also 3(d) below):

- the uncertainty of calibration of the DMM for the range used and
- the uncertainty in the use of the DMM arising from any drift in its calibration (see Note above) and AC pick up, etc.

2. The uncertainty in assuming the temperature of the thermocouple being calibrated is equal to that of the standard. In other words, the possible difference between the tip temperature of the thermocouple undergoing calibration and the temperature of the reference (temperature of the freezing metal, if a fixed point, or that of the standard thermocouple or platinum resistance thermometer, etc).

3. The uncertainty in measuring the signal produced by the thermocouple being calibrated. This will depend on whether the thermocouple is being calibrated alone or as part of a thermocouple/extension lead and/or instrument combination. If a combination is being calibrated the temperature at each connection (at the 'head', where thermocouple meets extension lead, and at the CJ end) must either be confined in some way or varied during calibration to estimate its influence (expressed as a corresponding component of uncertainty). For example, the head and/or CJ may be held at ambient temperature during calibration—this is desirable for a head because the extension lead then contributes little emf and thus error. The calibration report must stipulate the conditions applying to the head and/or CJ (at ambient or otherwise) and that the calibration data refer only to those conditions. In any case, we have the components:

(a) Inhomogeneity in Seebeck coefficient along the thermocouple being calibrated, to anticipate use with a different longitudinal temperature profile (that during calibration is unlikely to match the profile along the wires in a subsequent use).

(b) The effect of using extension leads (if any), either those accompanying the thermocouple or those added by the laboratory, as a temporary

4.7. CALIBRATION UNCERTAINTY

measure, if the thermocouple is considered too short for the calibration facility. The error contributed by an extension lead is proportional to the temperature interval it spans (from head to CJ). The head temperature may need to be measured during calibration so that it can be kept at ambient and to allow a measurement of the error contributed for a deviation from ambient of 10°C, say.

(c) The effect of using a non-zero CJ temperature. Usually the flexible section of thermocouple at the CJ end has been cold worked, through handling, and has not received the same degree of annealing as the alumina-insulated section. As a result its thermoelectric properties are different, and the calibrating laboratory may choose to have the CJ at ambient temperature to avoid a significant gradient in the region. If in a use subsequent to the calibration the CJ is placed at 0°C, the user should estimate the associated error and include it as a component of uncertainty in that use.

(d) An uncertainty to cover the instrument used to measure the thermocouple emf, unless it forms part of a thermocouple/instrument combination being calibrated. If the instrument (preferably a DMM or DVM) differs from that used to measure the thermocouple standard the following components must be included. Alternatively, if the same DMM is used for both thermocouples, the following are included only if the thermocouple emf is significantly different from that produced by the standard (see Note below):

- uncertainty of calibration of the DMM (from its most recent calibration report—the uncertainty may differ from calibration to calibration) and
- uncertainty in the use of the DMM arising from a drift in gain and/or zero since its calibration (see Note on page 138) and AC pick up, etc.

Note: if the thermocouple emf measured at each calibration temperature is nominally equal to that produced by the standard (i.e., the thermocouples are of the same type) and the same DMM/range is used for both, the non-linearity and gain errors of the DMM cancel, unless excessive, and their contributions should not be included in the calculation. The errors are 'correlated' (page 163) and there is then a need to include only type A (random) effects and zero errors. Even if the thermocouples are of different types, and the same instrument range is used, partial cancellation of the errors would occur—and the chosen uncertainty components could reflect this. However, this complication may be ignored in practice, and the uncertainty contributions be considered for each, because errors contributed by the DMM should be insignificant—it is easy to choose a DMM with sufficient accuracy and stability to have little effect on the calibration of a thermocouple.

4. Effects not considered above.

   (a) Scatter in the data relative to the best-fit curve (from a calibration plot like Figure 4.4). Such a plot highlights errors present in the calibration that would otherwise not be accounted for.

   (b) The error that may arise when interpolating between tabulated data to be given in the calibration report (for example, an uncertainty component of $10\,\mu V$ applies to Table 4.3—see Figure 4.4 at 50°C).

   (c) Rounding errors (section 5.5) when reporting data.

   (d) Contributions mentioned in item 18 on page 183.

Some uncertainties are not always relevant. For example, in an *in situ* calibration (section 4.10) the only errors that apply are those numbered 1 and 4 above. Whereas error 2 may not normally be included for an *in-situ* calibration, its possible variation from calibration to calibration should.

4.7.1 An example

To demonstrate how a calibration uncertainty would be calculated, let us calculate the 'best' that could be achieved for a new base-metal thermocouple intended for use in the temperature range 0 to 1100°C. Assume the standard is a type R thermocouple calibrated at the National Measurement Laboratory with an uncertainty that covers its use at any immersion depth. As a practical and accurate measurement set-up, consider the following (see also the section beginning on page 129):

- the tips of the two thermocouples are wire-wrapped together in the uniform-temperature region of a suitable tube furnace (such as that shown on page 128),

- each thermocouple is connected to extension wires (not 'compensating': see section 3.4), which meet copper wires in an automatic ice-point reference (0°C),

- and the copper circuit continues through a low-thermal switch to a DMM (on a suitable mV range).

The component uncertainties are given in Table 4.5, expressed in temperature units.

Errors (f) and (m) of Table 4.5 apply whether the calibration is calculated in emf or temperature units and, despite being zero in this example, were included for completeness and as a reminder. If the measuring range of the instrument had been set to read directly in temperature, rather than for emf, errors (f) and (m) could have been highly significant, because the

4.7. CALIBRATION UNCERTAINTY

two conversions, to and from temperature, would have been done by the instrument.

Notice that errors (d) and (k) tend to compensate each other. If both the standard and the thermocouple being calibrated were of the same type, the errors in assuming the CJ temperature was 0°C tend to cancel. Because the Seebeck coefficient of a type R thermocouple at 0°C is roughly half that at high temperatures the compensation is only partial. Thus, I have ignored error (d) and retained (k) as an over-estimate. In any case, their effect on the overall uncertainty is small for the conditions assumed.

Table 4.5: Component uncertainties for the calibration of a base-metal thermocouple (TC), calibrated against a type R standard (STD). The values are effectively components of expanded uncertainty combined using the method of page 170 (see Note on page 172).

	Error	Component Uncertainty	
(a)	calibration error (NML) of STD	0.1°C +	0.035% of T
(b)	error in use of STD other than (a) & (d)[†]	0	0.05
(c)	extension wires of STD[††]	0.1	0.01
(d)	knowledge of CJ temperature	0	0
(e)	instrum. error: measure STD	0.3*	0.01
(f)	convert V_{ref}^o to T[‡]	0	0
(g)	tip temperature difference	0	0.02
(h)	AC pick up problems	0	0.05
(i)	scatter of data, interpolation & rounding	0.1	0.02
(j)	extension wires of TC[††]	0.4**	0.03
(k)	knowledge of CJ temperature	0.05	-
(l)	instrum. error: measure TC	0.1	0.01
(m)	convert T to V_{ref}[‡]	0	0
(n)	inhomogeneity of TC	-	0.1
(p)	drift in TC during cal.	-	~0.1
	overall uncertainty (root sum of squares)***	0.54	0.167
	overall uncertainty (rounded up):	0.6°C +	0.17% of T
	overall uncertainty, if errors (c), (j) & (p) are avoided and DMM is zero corrected:	0.15°C +	0.13% of T

[†] drift component discussed on page 123.
[††] due to mismatch in emf for the conditions assumed in Table 4.1
[‡] use of reference tables - assumed done correctly.
* can eliminate by measuring circuit zero & correcting.
** can reduce by choosing a lead with negligible calibration error.
*** see comments on page 173 regarding the calculation of overall uncertainty for this table.

Notice also that there are 15 errors listed for this example. In practice, the estimating of these components should not be a problem because the final result, the overall uncertainty, will depend mainly on only a few components (see section 5.4). These must be assessed carefully and the others may be simply guesstimated! In Table 4.5 the only critical components are errors (n) and (p), provided (e) and (j) are reduced as indicated. Of course, if the measuring set-up differs from that described other errors may become critical.

4.8 Use of a calibration report

The following procedure is suggested when using a thermocouple calibration report containing a table of data in emf units, like Table 4.3 on page 135. A measurement is made using the calibrated thermocouple and its emf is corrected (if necessary) to a CJ temperature of 0°C to give the value V. From V, the calibration error for the thermocouple at this temperature (unknown at this stage) is $(V - V_{ref})$, which is obtained by linear interpolation of the calibration data (page 135). The subtraction of $(V - V_{ref})$ from V yields V_{ref} (see example below). Since V_{ref} is the reference-tables value corresponding to the tip temperature, T, the latter temperature may then be obtained by feeding V_{ref} into the reference equations/tables.

Many users have difficulty understanding the intermediate step involving $(V - V_{ref})$ needed to calculate T, so an example will now be given. The procedure for a calibration expressed in temperature units is similar (see Table 4.4).

Suppose the thermocouple represented in Table 4.3 (page 135) develops $5241\,\mu\text{V}$ for a CJ temperature of 0°C. Linear interpolation of the data in the table yields $61\,\mu\text{V}$ for $(V - V_{ref})$ at this emf. This is calculated as follows, noting that from 100 to 200°C, $(V - V_{ref})$ increases by $4\,\mu\text{V}$ from 60 to $64\,\mu\text{V}$ and V increases by $4042\,\mu\text{V}$ from $4096\,\mu\text{V}$. Thus at $5241\,\mu\text{V}$, $(V - V_{ref}) = 60 + 4 \times (5241 - 4096)/4042 = 61.1\,\mu\text{V}$.

Actually, there is little need for calculation in most cases. A sufficiently accurate value of $(V - V_{ref})$ may be obtained by inspection of the calibration table, such as Table 4.3, because the amount by which the reported values of $(V - V_{ref})$ increment is usually not large compared to the calibration uncertainty. In the above example, the calibration uncertainty is not likely to be less than $50\,\mu\text{V}$ and it is not necessary to estimate $(V - V_{ref})$ to better than a tenth of this, $5\,\mu\text{V}$. Thus, any value between 60 and $64\,\mu\text{V}$, the nearest values in Table 4.3, would do and $61\,\mu\text{V}$ looks about right for an emf near $5000\,\mu\text{V}$.

Therefore, with $V = 5241$ and $(V - V_{ref}) = 61$ the emf that would have been produced if the thermocouple had been ideal, i.e., in conformity with the

4.9. MOVABLE TEST PROBES

reference tables, is V_{ref}, where

$$\begin{aligned}V_{ref} &= V - (V - V_{ref}) \\ &= 5241 - 61 \\ &= 5180.\end{aligned}$$

From the reference tables for type K thermocouples, $5180\,\mu\text{V}$ corresponds to a temperature of 126.4°C, or 126°C when rounded.

4.9 Movable test probes

By 'movable test probe' I mean a thermocouple calibrated in the laboratory and used in the 'field' for short periods of time, totalling less than, say, 100 h. For example, one of a number of base-metal thermocouples used for assessing the temperature uniformity of heat treatment furnaces. They could be used once only, and discarded, although they are more usually re-used several times at the same immersion or at a different one.

For the data from movable test probes to be reliable, consideration should be given to limiting irreversible changes in Seebeck coefficient, stabilising reversible change and whether quenching is appropriate. These issues are discussed on pages 118 to 122.

In this section, test probes are singled out as a useful example of how calibration procedures may be tailored to a particular application: the testing of nominally uniform work spaces in furnaces and ovens, referred to as 'enclosure testing' (Chapter 6). For example, 12 thermocouples may be required and once installed will each have different immersion depths and be used for about 1 h at each of several temperatures. In choosing a calibration technique we need to account for the difference in calibration between the sections of thermocouple supplying emf during calibration and those sections producing emf in the enclosure test.

Test probes may be calibrated during the enclosure test, by having a reference standard present, or after the test, in the enclosure, with their tips together and their critical emf-producing zones at the positions they will occupy during the test. It is usually more practical, however, to calibrate them in the laboratory—at immersions short enough not to affect wire that will produce significant emf during the enclosure test.

Consider the following procedure.

a. Fabricate/select a batch of thermocouples cut consecutively from the same reel or coil, e.g., 12 probes each 5 m long.

b. Calibrate (sections 4.4 and 4.5) each probe at the shortest depth of immersion, say 200 mm, in a suitable furnace (for example, that shown on page 128). Take note, however, of item (8) on page 178.

c. Conduct the calibration at a number of convenient temperatures over the range of interest and obtain correction data for the specific temperatures of the enclosure test by interpolation (section 4.6). Alternatively, choose a sequence of temperatures and times at temperature similar to those to be encountered in the enclosure test.

d. Install the probes in the enclosure at immersion depths ~ 200 mm greater than that used in **a** above, although see Note below. This results in fresh, unused wire lying in the emf-producing zone at the wall of the oven/furnace.

e. Subsequently, if the probes are to be re-used at a different set of sites in the enclosure or in a different enclosure, the depths of immersion should then be increased a further 200 to 300 mm—see Note below.

Note: steps **b** and **c** will not be necessary if the required properties of the thermocouple coil are known from earlier calibrations. The requirements placed on immersion depth in steps **b**, **d** and **e** above are not necessary if the temperature of use is restricted to below that causing significant change in Seebeck coefficient—to below 200-250°C for as-new wire, for example. For further detail on temperature ranges for stable operation, depending on choice of thermocouple and its state of anneal, see section 4.3 on page 118. See also the comments on thermocouples for oven testing on page 203.

The relatively short (~ 200 mm long) sections of thermocouple supplying emf in the enclosure test have not been calibrated directly. Nevertheless, we can assume that the mean of the calibration data obtained from the 12 probes will apply to each when in the enclosure. Then, as an estimate of the error implied in this assumption, we use the spread of data among the thermocouples at each calibration temperature. The data are direct indications of the variation in Seebeck coefficient along the original coil of wire (or MIMS cable) sampled at 5 m intervals (for the example in **a** above). Thus, we take $1/2$ the range in the corrections at each temperature as the inhomogeneity component of uncertainty.

If the batch of probes are to be used with a data logger with ACJC (designed for thermocouple input and reads in °C) a different approach may be taken. A data logger has an array of input terminals and the temperature variation across them (maybe 0.3 to 1.0°C) introduces a further problem—the ACJC network, having a single sensor, cannot correctly compensate for all inputs. Fortunately, these CJ temperature errors are stable once the instrument has warmed up. I suggest, therefore, that the thermocouples and data logger be calibrated as an integrated unit, that the thermocouples be assigned specific input positions and that they be calibrated and subsequently used at these positions. Then calibration corrections should be determined and applied for each thermocouple/input combination and since the variation between them is not a measure of inhomogeneity (as in the above case), the latter must be taken as ±0.1% (base metals), unless previously assessed.

A sample calculation of the calibration uncertainty for test probes is given in section 4.7.

4.10 *In-situ* calibration

An *in-situ* calibration is not only the only meaningful form of calibration in many cases, but it has considerable practical value in all fixed-immersion situations. This is so because it relates to the particular temperature distribution experienced by the thermocouple, and thus to its unique distribution of thermoelectric change, and it includes the effects of all components of the measurement system, as well as the systematic errors (section 5.3) arising from the particular location and its surroundings. On the other hand, the laboratory calibrations of individual components yield more accurate values of their intrinsic properties. Laboratory calibrations, however, may be of little use in some cases, e.g., for compensating leads. Also, they may apply to inappropriate environmental conditions, such as temperature and humidity, and, naturally, they don't include the location-related systematic errors peculiar to the 'field' use.

The results of any one *in-situ* calibration refer only to that particular time and situation, and as the thermocouple and other components drift the change may be monitored by periodic *in-situ* calibrations. The level of change at intermediate times is determined by interpolating the data graphically, and extrapolation is reliable if the drift is well behaved.

It is usual for the reference thermocouple to be a laboratory-calibrated Pt-based working standard. Such a probe has rigid insulation and is usually mounted in a protective ceramic sheath. Thus, it may not be possible to position its tip near that of the resident thermocouple. Even if they are in close proximity the difference in their temperatures is likely to be significant. This is not serious, however, if it is safe to assume that the difference is systematic and stable. Then, an initial laboratory calibration (section 4.4) of the resident thermocouple can be used to correct its first *in-situ* calibration done soon after installation. Subsequent *in-situ* calibrations, with the working standard at the same site each time, would be used to monitor the relative changes.

Clearly, for an inflexible working standard, the weak link in an *in-situ* calibration is the difference in temperature between the tips of the resident thermocouple and the standard. Every attempt should be made to reduce this difference and, if possible, arrange for it to be stable and reproducible. If the two temperatures are equally representative of the process then all is satisfactory and the calibration is reliable, both in an absolute and in a relative sense. If not, the absolute calibration of the measuring system must be obtained by summing the individual laboratory calibrations of all important components. Then, periodic *in-situ* calibrations serve to indicate the (relative) drift.

Fortunately, there are alternatives to the Pt-based working standard. Calibrated, flexible base-metal thermocouples could be routed to the resident thermocouple and their tips placed or wired together, while the furnace is at ambient temperature. After calibration the reference thermocouple may be discarded, if of a bare-wire type, or kept for re-use if suitably annealed and of the ID-MIMS variety (see section 4.3). Alternatively, if the process temperature is not disturbed by the removal of the resident thermocouple, its calibration can be periodically checked by brief interchanges with a similar thermocouple, calibrated for the purpose.

The choice of equipment for the reference probe needs some thought. It is easy for the measurement uncertainty to build up to an impractical level [119]. For example, assuming a 95% probability level (see page 156), a type R reference thermocouple may have been calibrated to an uncertainty of (0.5°C+0.2% of T) and, if close to the resident thermocouple, their temperatures could be within ±0.1%. The use of compensating lead to span an interval of 40°C contributes a component uncertainty of 0.2% and a good quality 1°C-resolving instrument adds a further (2°C+0.15%), say. The overall uncertainty (root sum of squares: see section 5.4) in the test or calibration equipment amounts to (2°C + 0.34%), or 0.6% at about 700°C, say. This level of uncertainty is satisfactory for a thermocouple used for process control or monitoring, provided the conditioning specification for the process allows a temperature variation of ±3%, or more, in the material being processed. Then, the uncertainty in the knowledge of temperature, being less than $1/10$ of the allowed temperature range (= $2 \times 3\% = 6\%$) for the material, will not unduly affect the process. Clearly, a specified conditioning tolerance of ±1% in temperature, as applies to many critical processes, will require a lot of care when choosing both the test equipment and the resident temperature-indicating system.

For processes or equipment that operate at several temperatures *in-situ* calibrations are usually done only at one or two of the more critical temperatures. For other temperatures this information serves as a guide, although it can also be used to estimate relevant calibration corrections. This is done by interpolation and the procedure is similar to that in section 4.6.

4.11 The calibration of temperature-measuring instruments

Temperature-measuring instruments designed for use with thermocouples have three characteristics in particular (see also section 3.5 on page 89).

1. They have either a high-impedance input stage or, if requiring current, as do moving coil types, the external circuit resistance should be fixed and specified. To my knowledge all 'current detector' instruments for thermocouple use are clearly labelled with the required value of external resistance.

2. An ACJC network that operates on the assumption that the CJ of the thermocouple being measured is at its input terminals. The adjustment of this network is a suggested first step in calibration.

3. A means of converting emf into a value of temperature for the particular thermocouple type assumed. Usually an instrument would have adjustments for zero and gain and, once set correctly, the systematic errors in the emf-temperature relationship are all that remain. Indeed, the correct adjustment of the instrument and an estimate of these conversion errors (often referred to as the 'linearity error') are the reason for its calibration.

Broadly speaking, most instruments, regardless of the type of signal they process, are calibrated in two stages. The first stage is the adjustment of zero and gain, effectively the minimum and maximum of the range, and the second is the measurement of non-linearity—in this case, the departure from linearity of a plot of true versus indicated temperature. Actually, it is preferable to plot/examine the difference between the indicated and true values (see Table 4.6), which would be linear for a perfect instrument.

In addition, the 'dead zone' should be measured. The dead zone is the extent to which the input signal can be varied without producing a detectable response by the instrument. Its magnitude may vary over the scale and it occurs because of friction, wire spacing in slide-wires, inadequate gain or saturation of an amplifier stage. For a digital instrument the dead zone is taken as the least count, when there is no random noise and the amplifier is not saturated or malfunctioning—otherwise it is larger. Its value, if significant, should be communicated to the user, and it needs to be allowed for during calibration.

A thermocouple-measuring instrument may be calibrated alone or as an integrated unit with one or more dedicated thermocouples. What follows is a discussion of the former—the calibration of an integrated system is essentially

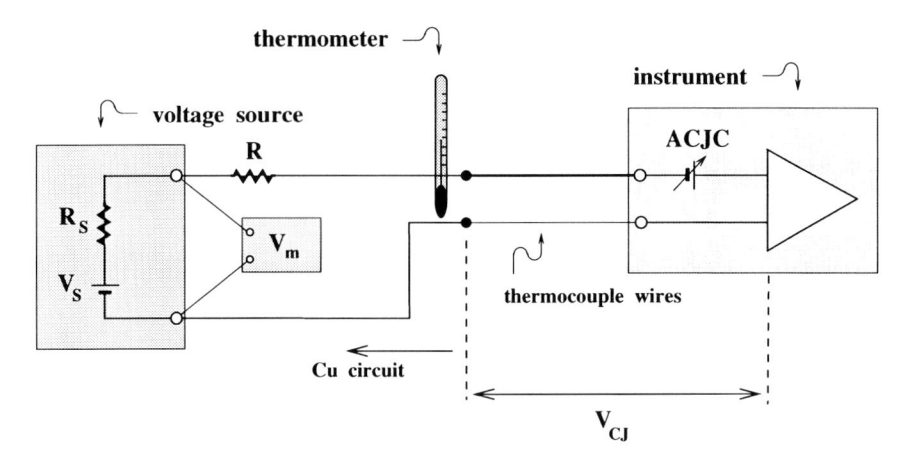

Figure 4.5: Circuit for calibrating an instrument with thermocouple input: using the 'thermometer method' (R is needed only if the instrument draws significant current and voltmeter V_m is needed if the instrument draws current and/or V_s is an uncalibrated source).

the calibration of a thermocouple (see section 4.5 and comments on page 144) with the above points kept in mind (and allowed for).

Zero emf applied to an instrument is equivalent to connecting a thermocouple whose tip is at the same temperature as that of its input terminals. Since the function of an instrument is to indicate (tip) temperature, it should indicate the temperature of the terminals, provided its ACJC system is correctly set. This argument is the basis for a commonly-used method of ACJC adjustment. A short circuit is placed across the input and the instrument is adjusted to indicate the temperature of a thermometer placed near the terminals. Unfortunately, the thermometer, at best, measures air temperature, which is not necessarily that of the terminals. The latter could easily be 10°C above ambient temperature because of heat generated within the instrument.

Instead, I suggest the input circuit of the instrument be extended to a convenient spot, more suited to having its temperature measured accurately. This is done by adding a length of thermocouple. The remote end of the thermocouple can be shorted and its temperature measured, using a thermometer, or fixed at 0°C in an ice pot. This allows accurate ACJC adjustment and forms the basis of the two methods of calibration described.

The first method, the **thermometer method**, is represented in Figure 4.5 and requires a calibrated thermometer to be in thermal equilibrium with a pair of junctions between the above-mentioned thermocouple extension wires and Cu leads to a calibrated emf source (used to represent the ideal thermocouple).

The second, the **ice-pot method**, is shown in Figure 4.6. An emf source

4.11. CALIBRATION OF INSTRUMENTS

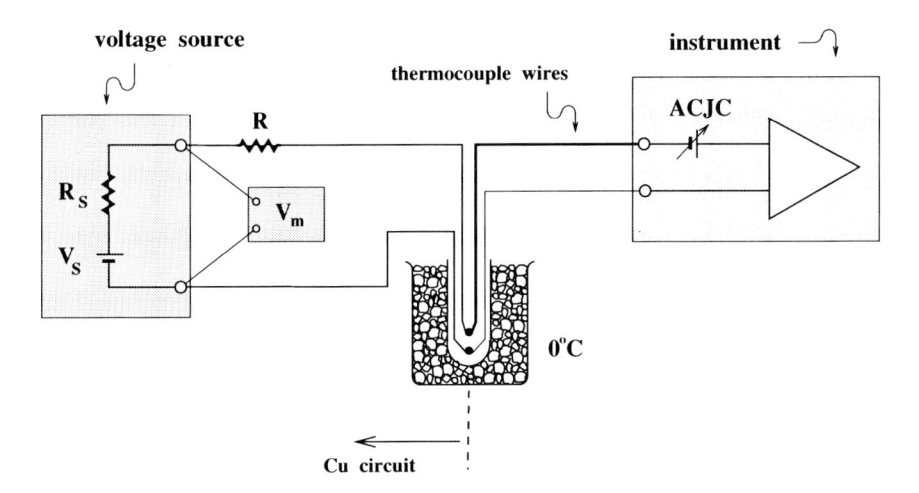

Figure 4.6: Circuit for calibrating an instrument with thermocouple input: using the 'ice-pot method' (R is needed only if the instrument draws significant current and voltmeter V_m is needed if the instrument draws current and/or V_s is an uncalibrated source).

of internal impedance R_s supplies a suitable signal, V_s, the resistor R is the external resistance required by some instruments (those with a significant input current, usually analogue types) and V_m is the emf measured on a calibrated voltmeter or potentiometer.

Thus, the circuit, attached to the simulated CJ, has an open-circuit output voltage of V_m and an impedance R. It therefore represents a thermocouple of impedance R and emf V_m, operating between an assumed tip temperature and that of the thermometer, in method 1, or 0°C, for method 2.

Procedure

As mentioned above, the input terminals of the instrument being calibrated are effectively extended, using suitable thermocouple wires, to a site at ambient temperature, and more easily measured (Figure 4.5), or at 0°C, and not requiring measurement (Figure 4.6). This temperature (that of the thermometer or 0°C) becomes the effective CJ temperature of the simulated thermocouple (V_m and R, in the figures) used to calibrate the 'instrument'.

So, the instrument calibration would have the following four steps. Although, to keep the description as simple as possible I have assumed the calibration errors of the 'thermocouple wires' are zero (there are two options for handling such errors—see the section on page 151).

1. Set the circuit to give $V_m = 0$ and adjust the ACJC network until

the instrument indicates the temperature of the thermometer, in the first method, or 0°C, in the second. Alternatively, for a suppressed zero instrument (usually with an analogue scale), set the voltage source to give a value of V_m equal to the reference tables emf for the lowest scale temperature minus that for the CJ temperature, as indicated on the thermometer or 0°C. Then adjust the ACJC network until the instrument displays its lowest scale value.

2. Adjust the gain, if applicable. This is done by increasing the signal until V_m equals the reference tables' emf corresponding to a suitably high instrument reading, with due regard to the CJ temperature (it may have changed with time), and setting the gain accordingly.

3. At roughly 10, 50 and 90% of the range measure the dead zone (P. 147).

4. Select maybe 5 to 10 equispaced values of temperature over the range of the instrument and at each do the following:

    - set the voltage source so that V_m is the reference tables' emf for a thermocouple having a tip at the chosen temperature and a CJ at the thermometer temperature or 0°C, whichever is applicable, and

    - record the instrument reading. If it is fluctuating choose an average value, and if the instrument relies on a mechanical operation (e.g., a moving-coil system), and has a significant dead zone, tap the case to bring the pointer to a central position.

Table 4.6 illustrates a method of recording data for the ice-pot method. The first two columns may be pre-printed, and in the ice-pot method, V_{ref} is the signal set up at the input of the instrument, as V_m.

On the other hand, the input emf required in the thermometer method is (V_{ref} - V_{CJ}) and must be calculated on the spot. The thermometer reading is taken, the equivalent emf, V_{CJ}, is obtained from the reference tables and is subtracted from V_{ref}. This is done for each calibration temperature and, clearly, there is a risk of operator error during the manipulations. The method is less accurate than the ice-pot method, it takes longer and requires more calculation.

The above information deals only with those common aspects of calibration relating to thermocouple emf. Some instruments have other requirements for their correct operation, such as damping adjustments and 'standardisation'. The relevant operator manuals should be consulted in these cases and their instructions followed.

4.11. CALIBRATION OF INSTRUMENTS

Table 4.6: Calibration data for a type K instrument, using the 'ice-pot' method.

Test Temperature T_o (°C)	V_{ref} (μV)	Instrument Reading T (°C)	Error $(T - T_o)$ (°C)
0	0	0	0
100	4 096	101	1
200	8 139	198	−2
-	-	-	-
-	-	-	-
-	-	-	-
900	37 326	904	4
1000	41 276	1001	1
-	-	-	-

ACJC adjustment

The thermometer method, described above, with the thermocouple wires formed into a tip at the thermometer, and without the calibration equipment depicted in Figure 4.5, is convenient for routine ACJC adjustments in the field. On the other hand, the ice-pot method, with the thermocouple wires formed into a tip and placed at 0°C, is also a convenient means of making the adjustments. It is also the more reliable method.

In each case, the instrument is adjusted until it displays the temperature of the thermocouple tip (that of the thermometer or 0°C), assuming the thermocouple calibration matches the reference tables (i.e., its calibration error, $E - E_{ref} = 0$). If the ACJC adjustment forms part of an instrument calibration, the thermocouple tip would not be formed. Instead, the thermocouple wires would be connected to a source set at $0\,\mu$V, and thus electrically equivalent to a 'tip'.

In general, however, thermocouple calibration errors are not zero and they need to be allowed for in one of two ways. Firstly, their direct effect on the adjustment of ACJC may be ignored. This is the preferred option, and it requires that the assessed errors for the wires be insignificant, i.e., less than about $1/3$ of the instrument uncertainty. For example, uncalibrated standard-grade wires could affect data in the ice-pot method by as much as ~ 0.6°C, if new (or ~ 0.3°C for premium-grade). For a smaller effect, a number of different wires may be calibrated (at say 20 and 50°C) and those with suitably small errors selected.

The second option is to apply corrections to cover their effect during the ACJC adjustment. This option is not preferred because of the potential for arithmetical mistakes. As an example, consider an adjustment based on a thermocouple whose tip is at 0°C. Its emf contribution to the instrument would then be negative (provided the instrument terminals are above 0°C!) and tend to cancel that from the ACJC network. The result is a zero signal presented at the internal DVM module (or the equivalent) and thus an instrument reading of 0°C, the correct indication for a thermocouple whose tip is at 0°C. Consequently, the effect on the instrument of any thermocouple calibration error would be opposite to that measured. For example, if the calibration error is +0.3°C (the thermocouple reads high) at 30°C, the presumed temperature of the input terminals, the instrument will display −0.3°C when correctly adjusted!

If thermocouple corrections are to be applied during an ACJC adjustment, or an instrument calibration, the temperature of the input terminals must be measured, to within a few °C (needed to interpolate the thermocouple calibration data). This may be done by first setting up the ACJC, without regard to the thermocouple errors, then short out the input terminals with copper wire. The displayed temperature is then that of the terminals.

For both options, a component of uncertainty covering the potential effect of the thermocouple wires on the ACJC adjustment must be included as a component in the uncertainty calculation for the calibration of the instrument or its subsequent use. In option one, the component is the maximum likely size (95% level—see page 166) of the thermocouple calibration error that was not applied, and, in the second, it is the maximum likely error in the correction that was applied.

Chapter 5

The Uncertainty in Temperature Measurement

5.1 Introduction

It is common to regard a measurement as **right until proven otherwise**. By 'right' I mean that every component of the measurement system is operating as it should (within specification) and that if everything looks OK, it is. Moreover, if there are any errors present—beyond those covered by the manufacturer's tolerances—they would be self evident, i.e., erratic or large enough to produce a ridiculous result. This view is simplistic, and numerous sources of error are overlooked. For example, a thermocouple assembly, its extension lead and the instrument may have been chosen with suitably small tolerances, totalling 5°C say, and correctly installed. All components may be functioning as they should and yet, the measurement may be in error by 50°C, because of influences not covered by the specifications. Three commonly overlooked sources of error are as follows.

1. The Seebeck emf in a thermocouple is smaller than other electrical signals in its vicinity, e.g., the emf equivalent to 1°C is at least 6×10^6 smaller (135 db) than the AC mains voltage. A good quality instrument would have the capability of rejecting, typically, only 60 db of normal-mode noise. AC interference may cause a large DC shift in reading.

2. Whereas the tolerances, or specified accuracies, stated for instruments are meant to apply for a reasonable period of time and, in any case, instrumental drift may be adjusted to zero at will, tolerances for thermocouples apply only to their unused state. Their calibrations will then drift, often to an unpredictable extent. Errors of 1 to 10% of temperature are common and even 50% is possible.

3. The measurement of thermocouple emf yields no more than the temperature of its tip. That there could be a difference between this temperature and that of the object of interest is often ignored. Also not realised, is that the presence of the thermocouple may have altered the temperature being measured.

The 'it's right until proven otherwise' approach is a natural partner to the practice of not supplying an estimate of uncertainty whenever a result of measurement is expressed. After all, the measurement is 'right'! Yet, as I will demonstrate in the following sections, a statement that 'the temperature was 480°C' is virtually useless unless accompanied by some indication of how accurate or reliable it is.

I suggest that a safer and more useful approach is to be continually conscious of the axiom that **every measurement is wrong!** It places the emphasis where attention is needed, and it leads naturally to the question of how wrong the result is and to a statement of uncertainty. Even if the latter is roughly estimated after a few moments thought, it nevertheless represents how good the measurement was thought to be at the time, and may prove invaluable on some future occasion.

This attitude better emphasises the true nature of measurement. In terms of effort, measurement should be an exercise in seeking out likely sources of error, in avoiding or minimising their effects and then in estimating the effect of those that remain. In other words, measurement is primarily a treatment of errors: obtaining the reading is easy.

Understanding how errors of various types can affect measurement is clearly a vital issue, but one thing needs clearing up first. An error is not simply a blunder or a mistake on the part of the operator or manufacturer—it has a wider meaning. The implications of error and the importance of estimating it are dealt with in the next section. In subsequent sections I describe the sorts of error that may contribute to a measurement, offer methods of reducing them and of estimating how much then remains and advise on the calculation of measurement uncertainty.

5.2 Error, accuracy or uncertainty?

Probably the most crucial reality in measurement is that the magnitude of a desired parameter, such as temperature or length, and its measured value are never the same. The difference may be small, but it may not. The unavoidable difference between the 'true' temperature and its measured value is the measurement **error**, and just as the true temperature is forever

5.2. ERROR, ACCURACY OR UNCERTAINTY?

unknown, so also is the error. All we have is our measurement—and it is wrong!

Clearly, what is required is some indication of how wrong the measurement is. Such a measure is the **uncertainty**. The word **accuracy** is often used instead, but erroneously—it is its inverse, effectively, and should be used only in the qualitative sense. Notice, for example, that a measurement done at an increased accuracy will have the smaller uncertainty!

Sometimes the uncertainty of measurement for a particular set of temperature-measuring equipment is carefully determined on one occasion and the value so obtained used for all subsequent measurements. In other words, the uncertainty is considered a characteristic of the measurement set up. This is incorrect!—other factors are involved. The uncertainty depends equally on each of the following:

- the quality of the equipment,
- the expertise of the operator and
- the intent of the measurement.

Obviously, the equipment sets a limit to the accuracy of measurement and, since measurement is effectively the art of handling errors, the expertise of the operator is paramount. But, just as important is the purpose of the measurement and, thus, how much care the measurement was thought to require at the time. For example, the temperature of an aluminium casting may have been measured because of the danger of melting (MP 660°C). The operator would distinguish between 500°C (no problem yet!) and 650°C (getting close!), would consider the values 490 and 517°C as being much the same and, possibly, 500°C would be entered in the record sheet for either. On this occasion, concerns about the ACJC adjustment or whether the equipment had an up-to-date calibration would not have been considered. On the other hand, the equipment may have the potential to be used to ±2°C, say.

Failure to give an estimate of uncertainty is a failure to complete the measurement. It is an omission that one day may prove irksome. A record of measurement, giving only the temperature of the casting as 510°C will not be helpful if, in 12 months time, the data is being re-assessed because of the in-service failure of a component previously adjacent to the aluminium casting. Was the measurement treated as non-critical, i.e., good to maybe 20°C (the uncertainty) and rounded to the nearest 10°C, or was the full potential of the system achieved, with an uncertainty of 2°C?

The best time to estimate the possible magnitude of error is at the time of measurement, and it need not involve lengthy calculation. In a rough measurement, for example, a quick off-the-cuff assessment may be all that is required, and the rough nature of the estimate should then be indicated

($\sim 20\,°C$, say).

So, (1) the temperature we seek is unknown, and remains so, (2) it is not sufficient to state only the measured value T_m, our best estimate of this temperature, and (3) an uncertainty must be included in the measurement result, as follows:
$$T = T_m \pm u.$$
But this isn't all there is to it. This simple statement, which sets boundaries for the 'true' temperature T cannot be taken literally. We cannot claim that T necessarily lies between the two specified values, $(T_m - u)$ and $(T_m + u)$, if only because u is merely an estimate and is not precise. Therefore, the statement is better read as: 'the temperature is **most likely** within the range $T_m \pm u\,°C$', but then what do we mean by 'most likely'? In other words, what is the probability that we have adequately covered the error?

Probability levels of 90, 95 and 99% are in common use, as is a measure for u often called the 'standard error' (see section 5.9), which implies a probability of about 65%. Often, $T_m \pm u$ is referred to as **confidence limits** (defined on page 187). This statement, when given at the 95% level, suggests that the observer had 95% confidence in claiming the temperature lies within the stated range and, conversely, that there is a 5% chance of it being outside the range.

It follows, that if the observer had chosen to have more or less confidence in the result, the uncertainty would have been correspondingly larger or smaller. For example, $476 \pm 5\,°C$ (99% confidence level) and $476 \pm 3\,°C$ (90% level) are roughly equivalent. But then, so also is $476 \pm 0.2\,°C$ ($\sim 10\%$ level), applying when only 10% confidence is chosen for the result. Clearly, the magnitude of uncertainty (u) is somewhat arbitrary, and for it to be properly interpreted the assumed probability level should also be stated. Unfortunately, it is often omitted, even in technical publications of good repute.

NOTE: in Australia, the 95% level has been adopted as the level of confidence (page 163) for its national measurement chain, which includes the CSIRO National Measurement Laboratory, the National Standards Commission and the National Association of Testing Authorities (NATA).

5.3 Error characterisation

It is useful to characterise an error in terms of how the particular source of error affects the measurement. An error is said to be **random** if its effect differs with each repeat of the measurement. On the other hand, a **systematic** error

5.3. ERROR CHARACTERISATION

does not fluctuate or change on taking repeats. It may drift with time, but as long as the measurement 'system' remains unchanged, the systematic error will be stable. By 'system' I mean all equipment and processes that contribute significantly to the measurement, including the operator.

The words 'random' and 'systematic' or their respective equivalents, 'type A' and 'type B' (see page 161), do not reflect on the nature of particular error sources. For example, thermocouple inhomogeneity will have a systematic effect, if measurements are repeatedly taken with the thermocouple fixed in position, and it contributes random error if each measurement is made at arbitrarily different depths of immersion. Parallax error in liquid-in-glass thermometry is most likely systematic if repeats are made by the same observer and, yet, it is random if different observers are used.

Random errors, as a group, are relatively easy to handle because they are 'visible', in the sense that they can be seen as the scatter in measurement data. We can reduce their effect by an obvious technique (see below) and we have ready-made formulae for calculating their contribution to the uncertainty.

Consider a digital voltmeter. It automatically 'takes' measurements regularly and frequently, and by observing the display the extent of random error is obvious. When asked to take a measurement, an operator will not usually read the first value focussed on. Instead, a sort of average of what is observed over a ~ 5 second interval will be taken—the 'average' is judged more stable and reliable, even by an unskilled operator. Further, the fluctuation evident in the repeated readings supplies useful information about the error source, and this can be used to calculate the corresponding uncertainty. To do this we need the techniques of statistics, given in section 5.9. Measurements are repeated, an arithmetic mean is taken as the best result and the uncertainty in this mean value, from random effects, is

$$m_p R, \tag{5.1}$$

where p is the percentage confidence level or probability, m_p is a factor given in Table 5.2 on page 188 and R is the range of measured values, the difference between the largest and smallest. The uncertainty is a component of the expanded uncertainty (see note on page 172).

Expression (5.1) is far simpler to calculate than that used in conventional treatments of error. My reason for proposing the different method is given in section 5.9.

Sometimes the aim of measurement is not to minimise the random variation but to examine its extent. Then, of course, expressions for the confidence limits should not be used. For example, a thermocouple is scanned (section 4.4) to examine the variation in Seebeck coefficient along its length as a guide to the possible systematic error applying to a future use at any one depth of immersion. Similarly, the calibrations of a number of thermocouples cut from the same supply reel may be used as an estimate of the 'inhomogeneity' error in any other thermocouple from the same reel. In these cases, the required uncertainty component is at least half the observed range, i.e., $0.5R$.

Let us now consider systematic errors. Unlike random errors, their effects are not 'visible' and cannot be disposed of by statistical means. There is nothing in the measurement to suggest their presence, let alone their magnitude, and this is where measurement 'art' and the expertise of the observer comes in.

The first step in handling such errors is to become aware of the possibilities—to seek out and identify the variety of sources that contribute error in a systematic way (see section 5.7 beginning on page 175). Having done this, they should be either removed or reduced and estimated. There are three methods of achieving a reduction.

Reduction by design Once a systematic source has been identified, the measurement system may be modified to reduce or avoid the problem. For example, the error in using a compensation lead may be reduced by reducing the temperature difference between head and CJ, or by replacing the lead with extension wires (head & extension wires are defined on page 85). Parallax errors in reading analogue meters may be reduced with an optical aid (telescope or mirror) to give a perpendicular line of sight. Also, a measurement system could include the option of using a different value of that parameter on which a particular error depends, e.g., in resistance thermometry the measuring current can be varied to estimate its heating effect [116]. Further examples are given in section 5.8.

Reduction by calibration Any stable systematic error that is characteristic of a component of the measurement system may be measured and then eliminated by applying a correction. The measurement of an intrinsic systematic error is known as calibration, and the application of a calibration correction replaces the uncertainty in the error, before calibration, with the calibration uncertainty, the uncertainty in its measurement. For example, the emf-temperature relationship of a new, un-calibrated thermocouple at 800 °C differs systematically from the reference tables. The manufacturers tolerance ($\pm 0.75\%$) may be treated as equivalent to an uncertainty of 6 °C with an associated probability level taken as $\sim 95\%$. A calibration may then yield a value -5 ± 1 °C for this difference and, so, on applying a correction of $+5$ °C the 'emf-temperature' uncertainty is reduced from 6 to 1 °C.

Reduction by randomisation A systematic factor or component in a mea-

5.4. TREATMENT OF UNCERTAINTIES

surement system may be replaced by a different, independent alternative and another measurement taken. If this is repeated with a different factor/component for each measurement the corresponding error will appear as a random contribution to the data. An average will then reduce the effect. For example, parallax error may be reduced by using a different observer for each reading and averaging; and an un-calibrated thermocouple can be replaced by different thermocouples, one for each reading and each from a different and independent coil.

Component uncertainties to cover the various systematic errors must be assessed before or during the measurement. This requires an investigation into the effects of each source or the reliance on experience gained in similar situations. The topic is discussed in section 5.7.

5.4 Treatment of uncertainties

Questions on how to treat systematic errors, of combining estimates of them to give an overall (systematic) uncertainty and whether to combine this 'total' with the cleaner, arithmetically derived confidence limits (section 5.9) for random errors has not been resolved in any absolute sense. Some authorities insist on keeping estimates of random and systematic error apart and some suggest that the combined uncertainty for systematic errors be expressed as upper limits, i.e., chosen so that the probability of the error lying outside the chosen limits is zero. The latter approach, requiring that the total systematic uncertainty be the arithmetic sum of the components, results in a somewhat larger value of uncertainty.

To resolve the issue, various international bodies, such as the International Organisation for Standardisation (ISO), are working towards international consensus on a procedure for dealing with uncertainty in measurement. Their most recent proposal [120], hereafter referred to as the ISO guide, is discussed in the next section, with particular attention given to its recommendations for terminology and methods of stating a measurement result.

One further point: the word 'uncertainty', when used qualitatively, refers generally to the doubt or lack of knowledge we have in a statement or a value, whereas, quantitatively, an 'uncertainty' is a <u>scalar</u> quantity that gives a measure of this dispersion or lack of knowledge. Similarly, the standard deviation s of equation (5.10) is always expressed as a scalar. If $\pm 6°C$ represents a region of doubt for a measurement of temperature, the number '6' is the uncertainty, not ± 6. In other words, an uncertainty of $6°C$ indicates that the error present during the measurement, which of course is unknown, is likely to have been within $\pm 6°C$ of the measured result.

A simpler method of determining uncertainties is described in the section beginning on page 170, a method that is essentially equivalent to the ISO recommendations and produces a similar result.

5.4.1 The ISO guide

The ISO 'Guide to the Expression of Uncertainty in Measurement' [120] deals with nomenclature and methods of evaluating, combining and reporting uncertainties. Its essential features are detailed below, and a summary is given on page 166. To begin with:

- The guide takes care to distinguish between the terms 'error' and 'uncertainty'—they are not synonyms, but represent completely different concepts (discussed in section 5.2).

- In general, the measured quantity, the **measurand** (e.g., temperature), is not measured directly and bears a functional relationship with other quantities (e.g., emf, resistance). The function may be quite complex.

For example, temperature may be obtained from a resistance thermometer using a polynomial to express its dependence on resistance. Errors in temperature may have had their effect via the same polynomial (i.e., resistance errors), some other function (thermals, quadrature-balance errors) or directly (e.g., gain error in the temperature indicator). Even if the functional dependence on measured quantities is complex the error in the measurand may be expressed as a sum of the effects of each error source on the measurand. Each term is simply the product of the (initiating) error and a sensitivity factor (usually the partial derivative of the above function). For example, the sensitivity factor for a thermocouple is the inverse of its Seebeck coefficient, it relates °C to μV, and for a 100 Ω PRT it is about 0.4 Ω/°C.

- All sources of error relevant to the measurement must be considered in turn and expressed as the potential effect <u>on the measurand</u>. Each error should be expressed as a component of uncertainty in the same units (°C, μV or Ω) using sensitivity factors, if required.

Generally, the raw data available for error-analysis[1] is in the form of:
- a confidence limit for each component of uncertainty arising from a random source ($m_p R$ is suggested, see below) or the standard deviation s (calculated

---
[1] In Australia, at least, there is a growing use of the term 'error budget', despite 'budget' meaning 'a list of expected income and spending for a future time', or some equivalent. The perfectly good word 'analysis' correctly expresses the process—a separation of something into its basic elements—and doesn't imply a future event.

5.4. TREATMENT OF UNCERTAINTIES

after a somewhat greater effort: page 187),
- calibration uncertainties for all calibrated components in the measurement chain—usually equivalent to confidence limits,
- 'max. uncertainties' and 'tolerances' for uncalibrated components, from manufacturers' specifications—usually expressed as implying (almost) the full range to be expected for the relevant error—and
- estimates for other error sources, most-easily estimated as 95% (or so) intervals, much like the claims of manufacturers. Methods of forming such estimates are discussed in section 5.4.4.

The ISO guide recommends, as a starting point, the expressing of all components as 'standard uncertainties' (see below). To do so requires a decision as to the type of uncertainty that applies.

- Uncertainties are categorised as either **type A** or **type B**, terms that merely reflect the different methods of evaluation. Type A uncertainties are those evaluated by statistical analysis (random errors—see discussion on page 186) and type-B uncertainties are evaluated by some 'other' means. The terms are introduced as a means of avoiding ambiguities that may arise when using the corresponding terms 'random' and 'systematic'. Such ambiguities do not occur if the approach given in section 5.3 is followed.

- A **standard uncertainty**, u, is the more general term for the 'standard deviation'—if the error involved is from a random effect the standard uncertainty is the standard deviation. Otherwise, for systematic effects, it has an equivalent meaning and is estimated from all available data/experiences, after making assumptions about probability distributions, probability level and degrees of freedom (see below). In each case, the 'variance' is the square of the standard uncertainty.

Usually there are few (often just one) type-A uncertainties involved in any one measurement—each represents the variability in any repeats of the measurement, e.g., as evident in the changing display of a DVM. Being visible, its effect is easily estimated: it is the standard deviation s, calculated from equation (5.10) (page 187). Moreover, the type-A uncertainty is usually the least significant component, and as such its estimate will have little effect on the final outcome. Thus, I suggest the more-easily calculated estimate $m_p R$ (on page 188) be used. Then, being a confidence limit, it is treated in the same way as the type-B uncertainties (see below).

The available information on type-B uncertainties (see above) is usually expressed in the form $\pm L$, essentially equivalent to confidence limits with no probability given (probably at least 95%). However, the ISO guide requires their conversion to standard uncertainties using a suitable reducing factor.

The easiest to deal with are the calibration uncertainties for any calibrated components in the measurement. Then, the reported uncertainty is either a standard uncertainty or, as is more usual, the equivalent of a confidence limit, the product of the standard uncertainty and the 'coverage factor' (defined below), from which the standard uncertainty may be calculated. The coverage factor should be given in the report, if not, assume 2.0 for a slight overestimate in the calculated standard uncertainty (up to 5%, see page 165).

In other cases, an assumption (a judgement and sometimes a 'guess') must be made as to the form of the population distribution from which each error derives. Then, the standard uncertainty is obtained by dividing L by a factor appropriate to the presumed distribution.

A rectangular distribution (factor = $\sqrt{3} = 1.73$) is the simplest and may be assumed if the error is thought to lie within known limits ($\pm L$) and is equally likely to be anywhere in the range. This is somewhat unrealistic, although it may be argued that a rounding error, which should be insignificant, is 'rectangular'. Generally, an error is more likely to lie near the centre of its distribution, and the simplest is the (hypothetical) triangular distribution, with a factor of $\sqrt{6} = 2.45$. For most error sources, however, the error populations are Gaussian (Normal), possibly with some truncation (manufactured items rejected if outside the tolerance). For example, histograms of calibration corrections for any class of temperature sensor or instrument will reveal the Gaussian shape, e.g., that shown in reference [114] for Pt-based thermocouples. The appropriate factor depends on what (unstated) portion of the Gaussian population the stated uncertainty, $\pm L$, represents—it is 1.96 or 2.58 if we assume 95 or 99% of the population, respectively, and less if the distribution is truncated (approaching rectangular).

Rarely is there sufficient information to choose other than the Gaussian distribution, as that most appropriate, or, for a manufacturer's tolerance, to decide on the coverage probability. Therefore, unless there is strong evidence to the contrary, I suggest the Gaussian distribution and a 95% probability be assumed. Then a reduction factor of 2.0 is required for L. To assume a manufacturer's specification has more than 95% coverage would require a larger value, but then the reasonable assumption that such a specification represents truncated data would lead to a compensating reduction in the factor. Notice that an error of 10% in the chosen factor is probably smaller than the 'error' in presuming its probability level (unknown) and in L, when guesstimated from experience. In other words, concern or argument about the 'right' choice for the distribution or its factor should be put in perspective.

So, in general, I suggest a reducing factor of 2.0 be used unless a different value is clearly indicated, such as that expressed in a calibration or measurement report.

5.4. TREATMENT OF UNCERTAINTIES

Once the various components of standard uncertainty are determined, they need to be combined.

- The overall effect of all (N) error sources that contribute to the measurement, both systematic and random, is calculated by taking the positive square root of the sum of the component variances. It is referred to as the **combined standard uncertainty** and is given the symbol u_c, where:

$$u_c^2 = \Sigma_{i=1}^N u_i^2 \tag{5.2}$$

and u_i is the standard uncertainty for the ith error source. It assumes all errors are independent and uncorrelated.

- Some errors may be correlated in their effect on the measurand. Then, the situation may become messy, with determinations of **correlation** coefficients and covariances [120]. In most cases, at least in conventional temperature measurement, the need for such complication may be avoided with care. For example, if a DVM is used (on the same range) to measure the emf from a standard thermocouple and that from a thermocouple being calibrated, the errors contributed by the DVM (zero and gain) are correlated. The DVM measurement uncertainties for both should not then be included in equation (5.2) because partial cancellation of the errors occurs. If the thermocouple emf's are similar in magnitude DVM errors will have little effect and component uncertainties are not required. If different DVM ranges were used the correlation may be slight. In any case, such errors (with a properly chosen DVM) will be insignificant, and a single component may be entered in the uncertainty-analysis table as a token and a reminder of the potential for error, in case a less accurate DVM is used on other occasions.

Just as the 'standard uncertainty' is the more general term for a standard deviation (above), the 'expanded uncertainty' is the more general term for a confidence limit (strictly speaking, applicable only to type-A errors).

- The **expanded uncertainty** U is defined as: $U = ku_c$, where k is the 'coverage factor' (see page 165). The expanded uncertainty is a more practical measure of uncertainty than the combined standard uncertainty because it provides an interval that is *highly likely* to contain the net effect of the errors present during measurement, rather than being merely a convenient measure of dispersion (like the standard deviation, which represents only ($\pm s =$) 68% of the population of possible data values).

Obviously, we need a value for the coverage factor k, typically 2 to 3 in magnitude, which depends on the chosen level of confidence (page 156)—the 'coverage probability'—and on the number of effective degrees of freedom for the measurement.

Generally, the number of 'degrees of freedom' is the number of independent variables (data) involved in a sum. This is easily understood for the evaluation of the standard deviation s (defined on page 187), since s is derived using the sum of n squares (of the n readings). However, for type-B sources of error it is difficult to imagine an appropriate sum to represent the component of uncertainty. Fortunately, there is a qualitative approach that produces a suitable value (see below). For the overall measurement result, an *effective* number of degrees of freedom is required to evaluate Student's distribution and, thus, obtain the expanded uncertainty.

- The number of **effective degrees of freedom**, ν_{eff}, for the overall measurement result is obtained from the Welch-Satterthwaite formula:

$$\nu_{eff} = \frac{u_c^4}{\sum_{i=1}^{N} \frac{u_i^4}{\nu_i}}, \qquad (5.3)$$

where ν_i is the number of degrees of freedom for the ith error source.

The standard uncertainty, u_i, appears in equation (5.3) to the fourth power. Thus, we can expect that some of the components will have little effect on ν_{eff}. As an instructive example, assume there are m significant components each with the same values of $u_i = u$ and $\nu_i = \nu$. From equation (5.3):

$$\nu_{eff} = m\nu. \qquad (5.4)$$

In general, the error analysis would be considered poorly done if the dominant uncertainty components have ν_i less than 8, i.e., are only roughly estimated. Assuming each of the dominant uncertainties have at least 8 degrees of freedom, we have that $\nu_{eff} \geq 8m$, and if there is more than one highly-significant component ($m = 2, 3 \ldots$) then $\nu_{eff} \geq 16$. Therefore, most measurements will yield a value of ν_{eff} in excess of 16, and allowing for all components, it would almost certainly exceed 10 (a value chosen for its significance later), unless of course the measurement is intentionally rough.

Consider now the determination of each ν_i, the number of degrees of freedom calculated or allocated for the ith component. For the type-A case (random effects), ν_i is simply $n-1$ for a standard deviation based on n repeat measurements. The number of independent terms in forming s is not n because the data are connected by one relationship: that used to calculate their mean.

When u_i is a type-B uncertainty, ν_i is calculated from a subjective assessment of the reliability (see below) of the guesstimated value for u_i:

$$\nu_i = \frac{1}{2}\left[\frac{\Delta u_i}{u_i}\right]^{-2}, \qquad (5.5)$$

5.4. TREATMENT OF UNCERTAINTIES

where Δu_i is the estimated uncertainty in the estimate u_i (i.e., the 'error in the error'!). If, for example, the relative uncertainty, $\Delta u_i/u_i$, is considered to be about 10%, 25% or as much as 50%, then the number of degrees of freedom is 50, 8 or 2, respectively, from equation (5.5).

We could, for convenience, make **qualitative** reference to these estimates as 'good', 'reasonable' or 'rough'. Then, for the initial estimate of each type-B component, L (see page 161), and thus of its corresponding value of standard uncertainty, a decision could be made as to whether it was a good, reasonable or rough estimate, and assign $\nu_i = 50$, 8 or 2, accordingly. In practice, it is unlikely that a finer distinction could be made, so these three levels would cover nearly all cases. Fortunately, the values chosen for ν_i are not particularly critical, as may be seen in the example in section 5.4.3.

The ISO guide [120] states that an estimate 'reliable' to 25% may be taken to mean the relative uncertainty is 25% and if 'reliable' to 50%, the uncertainty is 50%. In my view, this is an unfortunate choice of words—if the 'reliability' of an estimate increases the uncertainty should decrease! This is like using 'accuracy' for uncertainty. Thus, I suggest the use of 'reliability' in this (quantitative) context be avoided, and maybe the above qualitative system be employed instead (see Table 5.1).

- Finally, in calculating the **coverage factor** k, the Student's t-distribution is assumed, and for the 95% level of confidence k is taken as the value of t_{95} given in Table 5.2 (on page 188) for ν_{eff}.

The coverage factor k is 2.2 at $\nu_{eff} = 11$ (95% level) and reduces to 2.0 beyond $\nu_{eff} = 25$. Thus, k is within ±5% of 2.1 for all ν_{eff} greater than 10. As shown above, ν_{eff} will most likely exceed 10 for all but poorly considered measurements, and we conclude that k is typically 2.1 (within 5%, which is near enough, better than how well the component uncertainties are known).

The ISO guide recommends (its clause 7.2.3) that when giving a measurement result in terms of the expanded uncertainty U, rather than the combined standard uncertainty u_c, the following should be stated.

- the result, given in the form: $T = T_m \pm U$, with the units for both T_m and U indicated,
- the level of confidence, preferably 95% in Australia,
- the coverage factor k and, finally,
- using the premise that it is preferable to err on the side of too much information, give sufficient detail to enable the reader to re-calculate U. For example, list all uncertainty components and their methods of evaluation!

The guide is thorough and elaborate and as a result its full implementation is likely to be limited to the upper levels of the measurement/calibration chain, e.g., to the national standards laboratories (i.e., National Metrology Institutes, NMI's [109]). This applies especially to the last of the above requirements! Nevertheless, it would need to be understood to some extent by all users of calibrated equipment, at least to the extent of being aware of the above terms and, especially, of understanding the significance of the parameter U and the level of confidence (the coverage probability).

5.4.2 The ISO guide: a summary

The ISO guide defines a number of terms to be used in calculating a measurement result. Uncertainties are to be calculated and combined as follows:

- A list of component uncertainties should be compiled from available data—usually in a form equivalent to confidence limits (see page 161) and having a containment probability of (roughly) 95% or so. Each should represent the effect on the measurand and be in the same units.

- The effect of each error source should be expressed as a standard uncertainty, u_i, calculated from the available data using an appropriate reducing factor. The value 2.0 is suggested (page 162) unless otherwise indicated (e.g., when stated in a calibration report). A more complicated, 'rigorous' procedure may be followed, if desired, although it may be difficult to justify.

- The overall effect of all error sources is the combined standard uncertainty, u_c, calculated from u_i using equation (5.2).

- A preferred means of expressing the overall uncertainty is the expanded uncertainty, $U = k\,u_c$, where k is the coverage factor.

- To obtain k, we need the number of effective degrees of freedom, ν_{eff}, given by equation (5.3), which requires values of ν_i for the components.

- ν_i is either $\sim$20 for a calibrated item (or the value of ν_{eff} in its calibration report, if given) or, for other components, a value selected after deciding whether the estimate of uncertainty is good, reasonable or rough (see page 165).

- Finally, the results should be stated in the recommended form (page 165).

This summary, and the description of the guide beginning on page 160, assumes that significant errors are not strongly correlated. If this is not so, and the suggestion on page 163 cannot be adopted, the ISO guide should be referred to for further instruction.

5.4. TREATMENT OF UNCERTAINTIES

5.4.3 The ISO guide: an example

There are devices on the market known as dry-well or block calibrators. They consist of a temperature-controlled metallic block with at least one vertical well, typically 6 mm in diameter and 150 mm deep. Temperature control may be set to the nearest 0.1°C and has a short-term stability of this order, or better. They are suitable for calibrating temperature sensors, although thermocouples would need to be new or 'as-new', because thermoelectric scanning is not feasible.

To calibrate such devices Pt-resistance thermometers (PRT's) would be used. However, as an exercise in calculating uncertainties, and working with relatively large errors, consider using a type-K thermocouple directly connected to the input terminals of a good-quality digital 'thermometer', set for type K use and having a least count of 0.1°C. Furthermore, I shall limit the discussion to measurements at 200°C (the thermocouple may then be calibrated to a minimum of uncertainty and not experience any degradation in inhomogeneity with use).

The easiest way of compiling a list of potential errors is to proceed along a logical path, e.g., beginning at the bottom of the well. A more elaborate example of forming such a list begins on page 137. The following items detail the errors thought to be present in the measurement of temperature using the type-K thermocouple and its instrument—with the block controller displaying a stable value of 200.0°C during the measurement.

1. The short-term instability in the difference between the controller set point and the temperature at the base of the well should be included, if the calibration report for the dry-well calibrator is to be most useful. The instability arises from drift in the characteristics of the control sensor and of the controller. It is difficult to estimate. In this case, for a PRT sensor and good-quality electronics, it will not be significant compared to some of the larger components below. A token value of 0.05°C is chosen, and used as a reminder to include such a term in all similar calculations.

2. The difference in temperature between the block of metal and the thermocouple tip. It depends on whether the thermocouple contacts the bottom of the well, on the presence of oxide scale and other material and on thermal conduction losses up the thermocouple to ambient. A guide to the possible error may be had by moving the thermocouple around (allowing for inhomogeneity effects!) and by adding an extra thermal load (extra wires/rods beside the thermocouple). Assume an uncertainty of 0.1°C (i.e., after experimentation it was considered there was less than about a 5% chance of it being greater than this).

3. The calibration correction for the thermocouple was given as -1.7 ± 0.3°C at 200°C, expressed as an expanded uncertainty with a containment probability of 95%. If the calibration authority had fully implemented the ISO guide, values of ν_{eff} and/or k would have been included in the report. They usually aren't. However, ν_{eff} would tend to be high, say 20, and its value is not critical (see below). Moreover, I will assume $k = 2.0$, possibly giving a slight overestimate for u_i (up to $\sim 5\%$).

4. The calibration data relate to a different temperature profile (immersion etc.) than that existing along the thermocouple in its present use. This difference represents an uncertainty equal to the inhomogeneity level $\times \sqrt{2}$ (effectively two measurements), except that the two values are partially correlated (see page 163). Since the short-range and long-range contributions to inhomogeneity tend to be of similar significance, I suggest we roughly compensate for correlation (short-range effect in this case) by ignoring the $\sqrt{2}$, i.e., take the inhomogeneity component for the measurement as equal to the assumed level of inhomogeneity. If this component had been included in the calibration (it should have) then item 4 is not required. For this exercise, though, assume it had not been included and that the 'tolerance' on inhomogeneity for new wire given on page 13 ($\pm 0.1\%$ or 0.2°C at 200°C) applies. Then, the uncertainty component is 0.2°C.

    If the thermocouple had not been calibrated, and was of a premium grade, the above components 3 and 4 would be replaced by the manufacturer's tolerance (page 28: 0.375% of 200°C) equal to 0.75°C.

5. The calibration correction for the instrument was $+0.2 \pm 0.3$°C (expanded uncertainty at 95%) and, as is usually the case, no component for future drift was included. It was calibrated at an ambient temperature of 21°C, while set for 'external CJ', i.e., the ACJC circuit was not included (see item 7). As for the thermocouple calibration report, assume $\nu_{eff} = 20$ for this component and a coverage factor of 2.0.

6. It is now two months after calibration and a potential drift component of 0.1°C is estimated from the manufacturer's value for '90-day accuracy', etc.

7. The ACJC circuit was adjusted using a calibrated type-K thermocouple, kept just for this purpose, immersed in ice (0°C). The uncertainty in this adjustment was thought to be dictated mainly by the thermocouple calibration, i.e., 0.1°C. If the adjustment had not been done just prior to its present use, at a room temperature of 25°C, the temperature coefficient for the ACJC circuit (0.01°C/°C from the instrument spec's) would have contributed an additional uncertainty of 0.04°C.

5.4. TREATMENT OF UNCERTAINTIES

8. While measuring the thermocouple, the instrument display fluctuated between 201.6 and 201.7°C, i.e., over a range of $R = 0.1°C$, in 1 min, and the mean reading was taken as 201.65°C, the central value. Assuming the range was obtained for 20 readings, we have from Table 5.2 the value $m_p = 0.13$ and thus a type-A uncertainty of $m_p R = 0.13 \times 0.1 = 0.013°C$. Even if R had been larger, say 0.5°C, the 'mean' could still have been estimated in this way (by eye, while watching the flickering display) and the number of readings conveniently taken as 20 ($m_p = 0.13$). This appears rough, but it is all that is needed whenever the type-A 'reading' uncertainty is relatively small (see Table 5.1). If, instead, it was thought to be a significant component, the strict procedure of taking n readings and of calculating $\bar{x}$ and s (or $m_p R$) may be followed.

9. The effect of the instrument having a least count of 0.1°C needs to be included—i.e., 0.05°C (assumed to be a 95% estimate).

10. AC pickup and thermals: considered to be negligible, so use 0.05°C.

11. Since the controller has a least count of 0.1°C, the results given for the measurement should be rounded to this level (section 5.5).

The above components appear in Table 5.1, as values of L_i. They have been expressed in °C, since the device being calibrated indicates temperature. This required that errors in emf (thermals, thermocouple inhomogeneity, etc) be divided by $40\,\mu\text{V}\,\text{K}^{-1}$, the Seebeck coefficient. For convenience I have used 2.0 as the reducing factor (see page 162), equal to the coverage factor assumed for calibrations (items 3 and 5).

From equation (5.2) the combined standard uncertainty is

$$u_c = 0.255°C.$$

From equation (5.3) we obtain the number of effective degrees of freedom for the measurement result:

$$\nu_{eff} = 64,$$

where 96% of the denominator ($\Sigma u_i^4/\nu_i$) was due to items 3, 4 and 5. Let us now see how critical is the choice of the various ν_i, by replacing 20, that assumed for items 3 and 5, by 8. The result is

$$\nu_{eff} = 30.$$

From Table 5.2 the coverage factor is either

$$k = 2.01 \quad \text{or} \quad 2.04, \quad \text{resp.,}$$

essentially the same for both values of ν_{eff}.

Table 5.1: An uncertainty analysis for the example beginning on page 167: each component of uncertainty, L_i, a qualitative assessment of its reliability and thus the number of degrees of freedom, ν_i (from equation (5.5)).

Component ($i = 1 \cdots 11$)	Component Uncertainty L_i (°C)	u_i† (°C)	Type B Estimation	ν_i
1. control instability	0.05	0.025	rough	2
2. temperature diff.	0.1	0.05	reasonable	8
3. thermocouple cal.	0.3	0.15	—‡	20
4. thermocouple inhomo.	0.2	0.1	reasonable	8
5. instrument cal.	0.3	0.15	—‡	20
6. instrument drift	0.1	0.05	reasonable	8
7. ACJC	0.1	0.05	reasonable	8
8. readings	0.013	0.007	—	19††
9. least count (half)	0.05	0.025	good	50
10. AC pickup & thermals	0.05	0.025	reasonable	8
11. rounding	0.05	0.025	reasonable	8

† calculated using a reducing factor of 2.0 ($u_i = L/2.0$), equal to the coverage factor assumed for items 3 and 5 (calibrated).
†† $\nu_8 = 19$ because 20 readings were assumed, although, since u_8 is very small its value is not significant.
‡ these components of uncertainty are not necessarily type B and were not 'estimated', they were obtained from calibration reports.

Next, we calculate the expanded uncertainty, defined on page 163:

$$U = k\,u_c = 2.01 \times 0.255 = 0.513\,°C$$

The measured value (201.65°C) must be corrected for the calibrations of the thermocouple (−1.7°C) and its instrument (+0.2°C), otherwise use of their reported calibration uncertainties is invalid. Thus, the final measurement result is

$$\text{temperature} = 200.15 \pm 0.51\,°C$$

and refers to a 95% level of confidence. Of course, the values will need to be rounded to the nearest 0.1°C.

5.4.4 The low-fuss method

For those who do not need to follow the procedures in the ISO guide I suggest the following approach, referred to for convenience as the low-fuss method. It is the technique I have used for many years, until the appearance of the ISO guide.

5.4. TREATMENT OF UNCERTAINTIES

I have recently adopted the term 'expanded uncertainty' to be comparable to the ISO guide, and the method ends with a measurement statement that complies with the guide (in essence). The only difference is the method of obtaining a value for the expanded uncertainty, U. Both procedures involve some looseness. The low-fuss approach is somewhat loose at the outset, and as a result avoids any complication. By contrast, the ISO procedure is an attempt at rigour, but at various stages requires a variety of simplifying assumptions to evaluate the parameters. For example, in working with systematic error sources their probability distributions need to be presumed, to allow calculation of u_i, and the reliability of their chosen values of uncertainty needs to be assessed (subjectively) to obtain ν_i.

The method is far simpler than that of the ISO guide and would produce much the same value for the expanded uncertainty, when evaluated at the same level of confidence. Any difference there may be between the outcomes is likely to be no greater than the 'uncertainty' involved in estimating (or guessing) the effects of most systematic errors, in either procedure.

The starting point for the low-fuss method was the realisation that systematic errors usually out-number and out-size the random, and that the uncertainty components from random and systematic effects should be combined, if the measurement statement is to be of practical use. Central to the method is the maintenance of the error-containment probability throughout estimation and processing. The only difficulty in the procedure is the estimation of each systematic error—a difficulty also common to the ISO procedure.

The method requires components of uncertainty, U_i for each error source, estimated such that $\pm U_i$ is an interval containing 95% of the error population. In other words, U_i equals L_i of page 161 and in Table 5.1. Each uncertainty estimate U_i is evaluated/chosen by judgement using all available information on the variability of the error, including previous measurements, general knowledge on the behaviour of materials and instruments, manufacturer's specifications and calibration reports. The estimate should be such that $\pm U_i$ represent 'reasonable, practical' limits for the error, i.e., the error is 'unlikely' to fall outside the chosen range. By 'unlikely' I mean that the chance is less than 1 in 20, that each estimate has a containment probability of at least $\sim 95\%$. This definition of an uncertainty estimate is obviously very loose—it was chosen to reflect the uncertain nature of error estimation in most cases. In practice, though, such an approach causes little problem.

Sometimes there is insufficient local knowledge or experience to form a reasonable estimate for a particular source of systematic error. Then, an overestimate could be given (see item 16 on page 183). If, as a result, the expanded uncertainty for the measurement seems satisfactory (not too large), then all is well. Otherwise, more knowledge should be sought to enable a better, and hopefully smaller, estimate for that component (section 5.8). Alternatively, the systematic error may be reduced by one of the techniques mentioned above, beginning on page 158.

Conveniently, many of the components are available as 'maximum uncertainties' and 'tolerances' given by manufacturers, etc (see also page 161), and are directly suitable as estimates of component uncertainties at the $\sim 95\%$ level.

The effect of random error is also evaluated at this level, as 95% confidence limits, i.e., $\pm m_p R$ with $p = 95\%$ (see page 157) or expression (5.11), requiring the standard deviation. Then, having estimated U_i for each of the N sources of error, the expanded uncertainty becomes:

$$U = \sqrt{\Sigma_{i=1}^{N} U_i^2}. \qquad (5.6)$$

Each of the component uncertainties, U_i, is equivalent to an effective component of 'expanded uncertainty', as defined by the ISO guide (although the guide does not recognise such components).

The measurement statement expressing T_m as the measured value for temperature T should be a simplified version of that given above, in the ISO procedure:
- the result should appear in the form: $T = T_m \pm U$, with the units for both T_m and U indicated and
- the level of confidence stated, preferably 95% in Australia.

NOTE: unless otherwise expressed, the values of uncertainty given in this book are 'expanded uncertainties' (see page 163) determined at the 95% probability level. Moreover, all component uncertainties are effectively component expanded uncertainties (see above).

An example

Consider the example measurement in section 5.4.3, with the raw uncertainty data (values of L_i, and thus U_i) given in the second column of Table 5.1 (page 170). This is all we need.

From equation (5.6), the expanded uncertainty becomes:

$$U = \sqrt{(0.05^2 + 0.1^2 + 0.3^2 + \cdots + 0.05^2)} = 0.510 °C.$$

Hence, after rounding (section 5.5), the result of measurement could be expressed as: 'the temperature of the controlled block was $200.2 \pm 0.5 °C$ with

5.5. ROUNDING

a 95% level of confidence'. Essentially that obtained on page 170, but with much less fuss.

Another example of calculating uncertainties is given in section 4.7. Here, as is often the case, a number of errors tend to increase with temperature and the dependence is expressible in the form $U_\circ + U_T\%$ of temperature, T. Consequently, the expanded uncertainty U must also be a function of T. To find U, each component uncertainty must first be evaluated at various values of temperature over the range considered, and at each temperature U is calculated using equation (5.6). Then on plotting the resulting values, the functional dependence of U on temperature is obtained. For example, for the 14 components of uncertainty in Table 4.5 (page 141), the expanded uncertainty is 0.54, 0.82, 1.27, 1.74 and 2.06°C at the temperatures 0, 300, 600, 900 and 1100°C, respectively. These can be represented as $0.54 + 0.134\%$ of T, which when rounded becomes $0.6 + 0.14\%$ of T.

Sometimes, when preferring to over-estimate the expanded uncertainty, a simpler method of calculation may be used—that of treating the individual components, $U_\circ$ and U_T, separately. For example, the value of U_T for the expanded uncertainty is the root sum of squares of the individual values, $U_{T,i}$. In Table 4.5, mentioned above, this approach was taken, to yield $0.6 + 0.17\%$ of T, clearly an over-estimate.

5.5 Rounding

Numbers are rounded to remove unwarranted detail in the least significant digits. For example, if the measurement uncertainty is ±0.7°C then significance should not be placed on differences of less than 0.1°C, and certainly not at the 0.01°C level as implied by the value 472.18°C. In this case, the result should be rounded to 472.2°C.

Whenever a number is rounded the adjustment (0.02°C in the above) is an error, but it is not significant if done according to the recommended procedures, given below, and done only once. If rounding is applied at various stages in the manipulation or calculation of data the error can accumulate and become significant.

Rounding involves two steps, the rejection of redundant digits and an adjustment of the last digit retained. It is applied first to the uncertainty and then to the measured value or mean. As a guide, I suggest that the approximate nature of the chosen or calculated value of uncertainty be kept in mind and, therefore, that differences of less than $\sim 1/5$ of the uncertainty are not significant. The recommended rounding procedure is as follows.

- Round once only: when the final measurement result is being expressed.

- Wherever possible, express the uncertainty to 1 significant figure only, although 2 significant figures would be reasonable when the most significant figure is small, less than 3, say. Examples of the former are ±9, ±0.07 and ±0.04°C and of the latter, ±12 and ±0.25°C.

- Uncertainty values should be rounded up, e.g., 7.4 and 7.8 become 8, except where rounding down involves a change of only a few percent, e.g., 7.1 becomes 7.

- The measured or mean value of temperature should be expressed with its least significant digit being the last digit affected by the uncertainty. Depending on whether the uncertainty had been rounded to 1 or 2 significant figures this criterion results in the temperature value having a rounding interval of roughly $1/20$ to $1/3$ of the magnitude of uncertainty.

- The measured or mean value should be rounded, up or down, to the nearest digit in the last place retained except if the un-rounded value lies midway between the two alternative values. Then, choose the even round value.

More details on rounding procedures and definitions of terms such as 'rounding interval' and 'significant figures' are given in references [121, 122, 123].

5.6 Measurement: a summary

At this stage it is useful to summarise some of the important points so far covered in the chapter.

1. All measurements are wrong.

2. Measurement is essentially the art of handling errors.

3. A measurement result has two components: its value and its uncertainty.

4. Only a superficial knowledge of statistical formulae is required because random errors tend to be small.

5. Systematics are not self evident and, as a result, the uninitiated tend to have an unhealthy trust in measurement.

6. A 'root sum of squares' places more weight on the large components of uncertainty, the systematics, so more attention should be given to their reduction and estimation.

7. As a last step in stating the measurement result avoid redundant figures by rounding the stated value and its uncertainty.

5.7 Sources of systematic error

There are many ways in which a temperature measurement system can be affected by error and, as an aid to remembering and to dealing with them, it is useful to group them in some logical way. For thermocouples, I suggest the following groupings:

- changes in the temperature of interest due to the presence of the thermocouple,
- factors that prevent the thermocouple tip from being at the site of interest,
- intrinsic errors in the circuit,
- extraneous signals or influences and
- errors in signal processing.

Examples of individual systematic sources are given below, arranged in the above order.

(1) The object, whose temperature is to be measured, and its surroundings form a distributed three-dimensional thermal circuit. To insert a temperature sensor is to modify the circuit and, possibly, to change the temperature of interest. Let me illustrate this point with two examples (for a further example see page 185, where an effect of absorbed radiation is discussed).

Firstly, consider an object having a temperature higher than that of the surrounding gas. If a thermocouple is placed on the object it will affect the radiative and convective loss from the surface and, if the object is transparent to radiation, the thermocouple would absorb radiation that would otherwise not have affected the object.

Furthermore, where the (insulated) thermocouple wires first contact the surface, thermal conduction along them will cause local cooling of the object, especially if its surface has a low coefficient of thermal conductivity (oxide layer, paint, ceramic etc). After all (see Figure 5.1), the heat flowing along the temperature gradient zone of the thermocouple (near the surface) must be fed from the surface, and for the flow to pass through a thermal-resistance layer (oxide film, ...) there must be an increased temperature difference across it. Moreover, the difference (error) is increased if a significant contact resistance [103] exists between the sensor and surface. Therefore, thermocouple attachment should be so arranged that the cooling effect, which must occur, be located some distance from the tip, sufficiently far to prevent it affecting the temperature being measured—the thermocouple wires (insulated) should run along the surface a suitable distance before leaving it (Figure 5.1). The optimum length of thermocouple contacting the surface is usually estimated by trial and error.

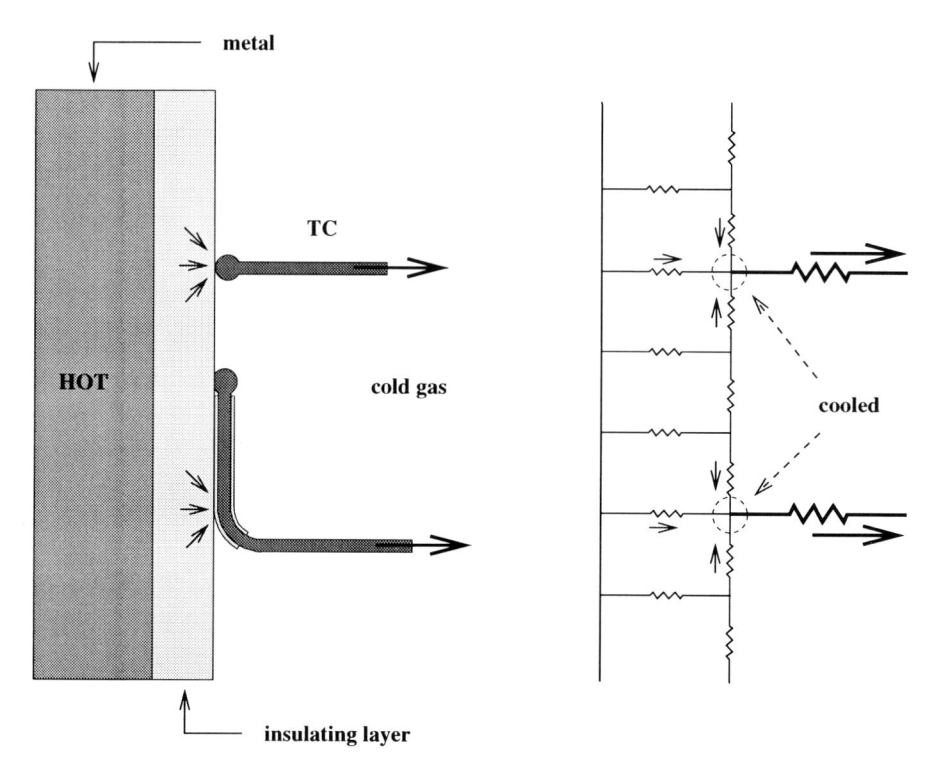

Figure 5.1: Thermal conduction paths, shown on the left, for two thermocouples placed on an object whose temperature exceeds that of the surrounding gas. On the right is the electrical equivalent circuit.

The problems of surface temperature measurement are discussed in reference [103] and an analysis of heat transfer for thermocouples contacting a hot surface is given in reference [105]. One solution is the use of a 'tapping thermocouple' [103] whose temperature is adjusted by an internal heater until no change in measured temperature occurs on tapping the surface. Then, no heat exchange can take place, while the probe contacts the surface, and no difference can occur across the surface layer (and/or any contact resistance).

Secondly, consider the measurement of gas temperatures, discussed in detail in section 3.10. Here, the main problem is that if any convective heat transfer with the probe tip is required at equilibrium, the temperature of the probe will not equal that of the gas. For example, a net flow of heat away from the tip by conduction within the probe and/or by radiative interchange with

5.7. SOURCES OF SYSTEMATIC ERROR

the surrounding surfaces (walls) will result in the tip being in a local pocket of cooled gas and the measured temperature will be low.

In both cases, thermal conduction along the wires to or from the thermocouple tip led to a change in tip temperature, because of a significant thermal resistance between the tip and the site of interest. This error is often referred to as the 'conduction error'.

(2) The thermocouple tip may not be at the site of interest and its temperature will differ as a consequence. Often, the preferred site is inaccessible because of the design of furnace or because the thermocouple needs to be surrounded by protective sheaths or radiation shields. Alternatively, the optimum site may be unknown—the intention may be to measure a representative temperature for a large region whose temperature distribution is not known. Also, if the temperature is varying with time the sensor may have an inappropriate response time. For the simple case of a sensor, of response time θ, placed in a fluid, whose temperature is increasing linearly at the rate dT/dt, steady state conditions would occur within say 2θ to 3θ. Then, the sensor temperature will have a constant lag of $\theta dT/dt$. For example, at $10°C\,min^{-1}$ and $\theta = 30\,s$ the lag is $5°C$. If the temperature is cycling the amplitude of cycle detected by the sensor will also depend on the response time.

(3) The emf-temperature relationship of that section of the thermocouple in the main temperature-gradient zone is assumed to match that in the reference tables. Alternatively, it may be assumed equal to that obtained from calibration using a different immersion and a different temperature profile. Both assumptions are subject to error. The various processes that cause change in a thermocouple's signature are dealt with in Chapter 2.

(4) The use of compensation or extension wires necessarily causes additional error. The manufacturer's tolerances may be large (if compensation cable, see page 85) and their calibration is of little direct use—corrections are not usually applied because of the unknown temperature interval spanned by the lead. There are also the more dramatic errors that arise from using the incorrect type of lead or when it is connected in the wrong polarity.

(5) There may be regions of the circuit that span a temperature difference yet don't generate an appropriate emf. Examples are connector blocks and switches with significant in-to-out temperature differences.

(6) The thermocouple and its extension lead should be the only sources of thermoelectric emf in the circuit. This is not always the case—temperature gradients in other portions of the circuit (across connector blocks, switches and the wires from CJ to amplifier) will also generate emf, known as 'thermals'. The hope is that each such extraneous source is balanced by a like source in the other leg. In any case, a two pronged attack will reduce them—where possible, avoid temperature gradients at terminals and switches and avoid dissimilar metals within them (copper, silver and gold may be together: they have small and similar Seebeck coefficients).

(7) The CJ of a thermocouple may not be at the temperature assumed for it and the temperatures of its two connections may differ (see page 12). Its measurement may be incorrect or, if relying on the known temperature of a uniform bath, the wires may not be sufficiently immersed. A properly prepared ice pot is the ideal CJ environment, provided, of course, the wires achieve thermal equilibrium with the ice/water mixture. For a description of an ice pot and some suitable dimensions see section 3.3.

(8) Binders, resins and/or water-proofing compounds used in the woven, flexible insulation of thermocouples tend to break down and boil off, especially above 200°C, and may condense in the temperature-gradient region when not given room to escape. Further, the organic materials may produce carbon, instead of carbon dioxide, if there is poor access to oxygen. These contaminants are likely to cause local changes in Seebeck coefficient. For example, a steady fall of 2 to $3°C\,h^{-1}$, for a tip temperature of 900°C, was noted in a study at CSIRO. This result was for 0.5 mm bare-wire type K thermocouples covered in a woven-silica material, suitable for temperatures up to 1000°C, and used in a loose-fitting, but nonetheless, constricting silica sheath.

For accurate work, particularly if the immersion is to change, these compounds should be removed, e.g., by heating the thermocouple at 300 to 400°C for an hour or so in an open space (centre of oven); and the resultant uniform, reversible change in Seebeck coefficient (see for example page 45) is allowed for by calibrating the thermocouples afterwards. For example, this decontaminating treatment is likely to change the emf of type K thermocouples by about +0.5%, depending on time and temperature. Sometimes, the treatment may be given more conveniently *in situ*, by first installing the thermocouple at an immersion 300 to 500 mm greater than required, kept in this position at > 300°C for one or more hours, depending on the temperature, and then withdrawn to the position chosen for its use. This ensures that the emf-producing zone is free of the above compounds and if done above 600°C, free of the hysteresis growth as well. Naturally, an allowance should be made for any change in calibration caused by the treatment.

(9) The electrical resistivity of the insulation between the wires may be low enough to affect the measurement. With quality materials and satisfactory fabrication techniques, the presence of an extensive 'hot spot' between the tip and CJ is usually necessary before insulation leakage becomes a problem. This point is illustrated in Figure 5.2, where the lowered insulation resistance for the hot spot, R, allows current and thus a potential drop across the loop

5.7. SOURCES OF SYSTEMATIC ERROR

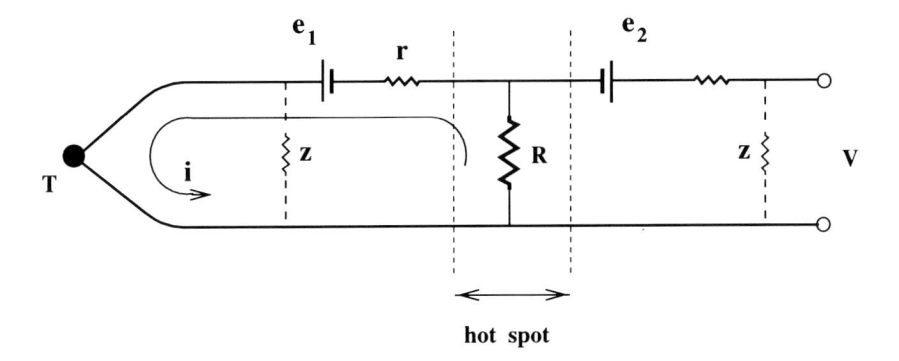

Figure 5.2: Thermocouple with a hot spot causing a lowered insulation resistance R (elsewhere Z too large for significant leakage) and a net current i. The Seebeck emf's are e_1 and e_2, tip temperature is T and the net output, V.

resistance, r. The current is $i = e_1/(r+R)$ and the signal measured across the thermocouple is

$$V = e_2 - e_1 + \left(\frac{r}{r+R}\right)e_1.$$

The desired result, that corresponding to temperature T, is $e_2 - e_1$ and the error is thus $r.e_1/(r+R)$, which becomes

$$\text{error} \approx r(\frac{1}{R})e_1$$

when R is large compared to r. Therefore, as a guide, the error is proportional to the length of thermocouple beyond the hot spot ($\propto r$), the width of the hot spot region ($\propto 1/R$) and the height of the temperature peak ($\propto e_1$). Clearly, the smaller the diameters of the components the greater is the effect.

Normally, the error is not significant when using the recommended ceramic materials (section 2.10), such as re-crystallised alumina with Pt-based thermocouples, for temperatures up to ~1600°C, and high-quality magnesia for MIMS thermocouples up to maybe 1200°C. In the latter case, however, there may be serious error if the ingress of moisture into the ceramic had occured during fabrication.

(10) Electro-chemical emf will be generated between the thermocouple wires if an ionic solution is present, e.g., when fibrous thermocouple insulation is wet. As indicated in Figure 5.3, the dissimilar thermocouple wires act as electrodes and the ionic solution produces a large DC chemical emf, E_0, and thus a circulating current around the thermocouple via its tip. The current is $i = (E_0 - e_1)/(r + R)$ and the measured signal, V, is in error by the potential

180   CHAPTER 5. UNCERTAINTY

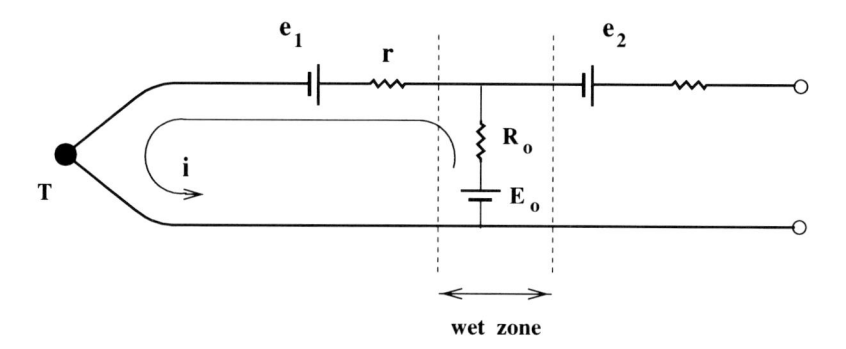

Figure 5.3: Thermocouple with part of its length wet with an ionic solution, which caused an electro-chemical emf, E_0, of internal impedance, R_0. Current, i, flows through the thermocouple loop of resistance r and the Seebeck emf's are e_1 and e_2.

difference across the loop resistance, r, i.e.,

$$\begin{aligned} \text{error} &= \frac{r}{r+R}(E_0 - e_1) \\ &\approx r\left(\frac{1}{R}\right)E_0, \end{aligned}$$

since, in practice, E_0 would be of order 1 V and thus much larger than e_1, and we may assume $r \ll R$, for simplicity. So, the error is roughly proportional to the wetted length ($\propto 1/R$) and to the length of thermocouple between the tip and the wetted section ($\propto r$).

Moreover, the impedance of the wetted section, R_0, will diminish the contribution of e_1 to the measurement—just as the hot-spot impedance R of Figure 5.2 did in (9), above.

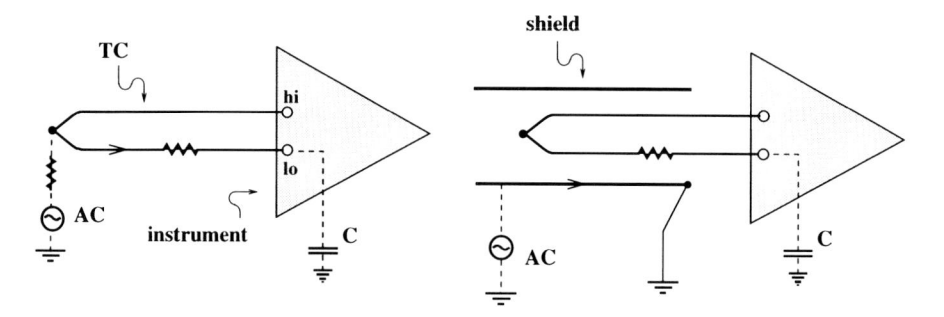

Figure 5.4: The effect of resistive and/or capacitive AC leakage from a furnace to a thermocouple (TC), on the left, and the result of intercepting the current with an earthed shield, on the right. The capacitance, C, represents internal leakage to ground, within the measuring instrument.

5.7. SOURCES OF SYSTEMATIC ERROR

(11) Leakage from DC or AC electric sources may be a problem. Any leakage current to the thermocouple will flow to earth via that thermocouple leg connected to the low side of the instrument input and thus generate a series-mode voltage proportional to the resistance of the leg. As indicated in Figure 5.4, the leakage current can be intercepted with a shield at earth potential. Connecting the shield to earth, however, may cause sufficient leakage current to trip an earth-leakage protection circuit, if connected. Then, the alternative of connecting the shield to a guard terminal (see Figure 5.5) may avoid the problem.

A guard terminal, if applicable, may be externally connected in a variety of ways when attempting to overcome a leakage problem. Two options for guarding are shown in Figure 5.5. One is to drive a shield and the other is a third-wire connection to the thermocouple tip. The latter option may not be effective, however, especially when the representation of leakage as a single

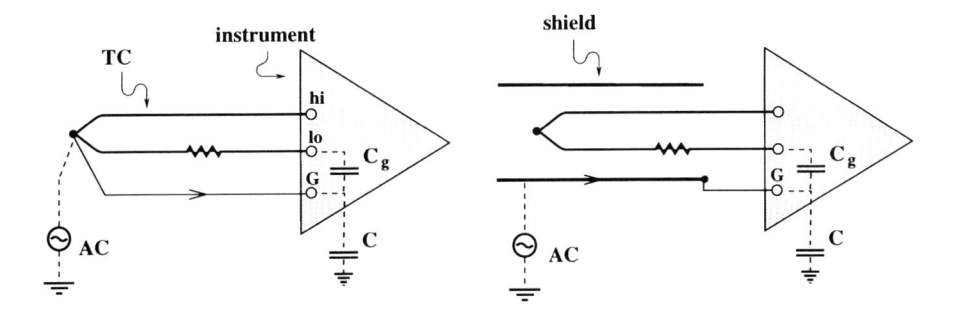

Figure 5.5: Electrical leakage (AC) diverted by using a guard (G) either to drive a shield, on the right, or connected by a third-wire to the thermocouple (TC) tip, on the left. The capacitance, C, represents internal leakage to ground and C_g, that between the input and its surrounding guard.

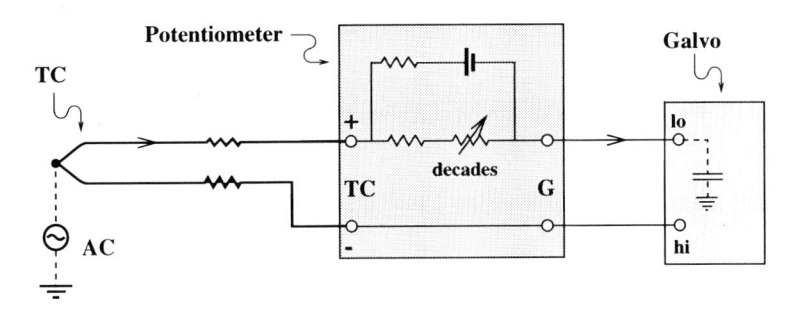

Figure 5.6: The effect of having the 'lo' terminal of a galvanometer connected to the high impedance side of a measuring potentiometer: AC leakage to the thermocouple (TC) causes a large AC potential difference.

path to the tip is an over simplification—leakage being distributed along the thermocouple length.

The guard, internally connected to a shield that surrounds the input stage of amplification, is internally driven to the 'same' potential as that of the input and as a consequence 'no' common-mode current can pass through the input circuit to the guard, i.e., the impedance of C_g (Figure 5.5) is effectively very high ($C_g \ll C$).

Some DC potentiometers are internally wired with the resistors of each dial in series with the 'lo' terminal of the galvanometer (Figure 5.6). If the latter is an electronic device, leakage from thermocouple tip to ground via the 'lo' terminal will pass through this resistance, often with large effect. In such cases, the connections to the galvanometer should be reversed.

Electronic instruments tend to respond to AC signals by both a shift in mean reading and an increase in their fluctuations. In other words, AC signals from any source will, in general, create both systematic and random responses. Of more concern, however, is that on some occasions—because of instrument design or the nature of the AC signal—the response to AC pick-up is only a DC shift—there being no significant increase in fluctuation/noise. Thus, there is then no warning sign that AC may be having a significant affect on measurement. Furthermore, AC currents may experience partial rectification at dirty contacts or as they traverse furnace insulation, etc. The instrument cannot distinguish between this DC component and thermocouple emf.

The presence of significant electrical interference may be checked by momentarily turning off the offending source—its effect on an instrument will be instantaneous and thus distinguishable from the thermal effect of any brief power loss (relatively slow).

(12) Unshielded portions of the circuit may act as an antenna for RF signals. Such high-frequency 'noise' could saturate the amplifier or be partially rectified at poor solder joints, for example, to give a DC affect. Spikes may upset the operation of electronics and if the received RF power is sufficient, significant heating of the thermocouple or other components may occur.

(13) Magnetic fields influence thermocouple measurements in a variety of ways. Common to any circuit is the electric current induced by an AC magnetic field if the area between the wires is non-zero and faces the direction of the field. It is avoided by running the circuit in twisted pairs. Other magnetic effects are related to the permeability of the materials and hence cause greater responses in ferro-magnetic metals such as iron and Alumel. If large enough, magnetic fields parallel to the axis of the wires (longitudinal) will cause significant changes in Seebeck coefficient, and if the field varies with time, magnetic hysteresis may lead to voltage spikes being generated. Also, transverse magnetic fields will generate unwanted signals if a transverse

5.7. SOURCES OF SYSTEMATIC ERROR

temperature gradient is present.

(**14**) There are numerous ways in which the instrument can supply error. ACJC, amplification, 'linearisation' and display are all processes that involve error. The dependence of instrument operation on ambient temperature requires some thought as well—an instrument is usually calibrated in the laboratory at only one value of ambient temperature and a systematic error results if it is then used at a significantly different temperature.

(**15**) The human operator may contribute error when extracting the indicated or displayed information and then in transcribing it. Rounding errors may also be significant.

(**16**) In calculating the expanded uncertainty, a list of error sources must first be formed, such as that on page 141 or beginning on page 167. Then, estimates are made of the corresponding uncertainties, each chosen to cover the likely range for the error. Whereas, for many error sources there is readily-available information to draw on, such as manufacturer's specifications, for others there isn't. Then, the experience of the operator is important.

Due to inexperience the estimate for any one component may be poor—significantly higher or lower than what an experienced person would have chosen for the same conditions. My suggestion here for the inexperienced operator is to seek further information and, as a last resort, simply over-estimate the component. In my view, the act of over-estimating is a legitimate way of including an additional uncertainty component to cover inexperience. This is equivalent to choosing a low value for the effective degrees of freedom, as would otherwise be done if following the ISO procedure (see page 163).

(**17**) With the increasing use of computers in temperature measurement, many of the tasks normally done by the instrument (ACJC and emf-to-temperature conversions) and the operator (calculations and rounding) are handled by software. It is imperative that the procedures built in to these programs are thoroughly checked, if possible by people not involved in their writing. Because of the power and speed of modern computers, more complex calculations and decisions are being asked of them and there is a greater chance of systematic errors escaping detection (via some obscure path in the program).

(**18**) The final measurement statement, containing the value being reported and its uncertainty, must be looked at carefully to see if all the assumptions that are involved are covered by components of uncertainty. For example, if it is claimed that the temperature indicated by an instrument is 507 ± 1°C, then in estimating the uncertainty, 1°C, only the component of item (15) above should have been considered. It should not include contributions from the instrument or the sensor. On the other hand, to report that the furnace temperature was 497 ± 7°C is to include far more assumptions and, thus, the need for more contributions to uncertainty. Possibly the most difficult assumption to cover is that the sensor temperature, which is all that is 'measured', is equal to the 'furnace-temperature'. To do this, a definition of furnace-temperature is required, after all, temperature in a furnace is not unique—it has as many values as there are sites (infinite) and at each it depends on the thermal properties of the sensor (section 3.10). A suitable choice may be some sort of average, or simply the temperature of a defined sensor at a specified site (see also Chapter 6).

In a similar way, the user of a calibration service should examine the calibration statement to see what is covered by the claimed uncertainty, assuming of course that the calibration laboratory was aware of their responsibility in this regard. For example, does the calibration report deal only with the characteristics of the sensor at 'the time of test' or does it also cover a period of subsequent use?

5.8 Experimental assessment of systematic error

When the systematic error from any one source cannot be avoided, or reliably reduced to insignificance, it needs to be estimated. Usually, this can be done by drawing on past experience or otherwise available information. Sometimes this is not possible. Then, it is advisable to estimate the error by experiment, e.g., by adopting the following, general procedure.

1. Examine the physics of the processes that generate the error.

2. Identify an adjustable parameter, X, that governs the magnitude of the error.

3. Measure a stable temperature with two or more measuring configurations having different values of X.

4. From the measured dependence of apparent temperature on X determine the level of systematic error, u_X, appropriate to the most practical measurement configuration.

5. The measurements of step 4 may be used to select an allowed upper limit for X and thus a maximum value for u_X. Provided X is kept within this range the uncertainty component associated with it can be taken as $\pm u_X$ when using similar probes in similar environments.

6. Alternatively, future measurements made under similar conditions may be corrected by $-u_X$ and the component uncertainty associated with the error source 'X' is thereby reduced to the uncertainty in this correction. The latter is calculated from the information obtained in step 4.

5.8. ASSESSMENT OF SYSTEMATICS

For example, consider a high-intensity lamp with a glass lens or window. The lens would be hot because of radiative power absorbed from the transmitted beam, convection through the gas behind the lens and conduction from the mounting. If the lamp is used in a coal mine the temperature of the outer lens surface is critical because coal dust falling on it could spontaneously ignite. But, the measurement of this temperature is not easy.

The temperature of a thermocouple placed on the lens surface would be systematically high because of the additional energy absorbed from the beam, somewhat more than that absorbed by the area of glass it covers. Also, the temperature is affected by the changed heat loss at the surface— the thermocouple would have changed the convective parameters and the emissivity of its outer surface differs from that of glass.

Let us examine that error arising from the additional energy absorbed from the beam. The rise in temperature, ΔT, would be roughly proportional to the absorptivity, and thus the effective emissivity ε, of the inner thermocouple surface facing the beam,

$$\text{i.e.} \quad \Delta T = \varepsilon F$$

where F is some function that relates to other physical characteristics of the measurement set-up. Also, as mentioned above, the measured temperature T' is also influenced by the changed heat loss from the outer surface, by Δ, say. So,

$$T' = T + \Delta T + \Delta,$$

where T is the 'true' temperature, the temperature of the glass surface if no thermocouple had been present.

To estimate F, and thus ΔT, measurements of surface temperature could be made with two very different and known values of ε. For example, fine thermocouple wires could be mounted on small squares of aluminium foil, one shiny ($\varepsilon_1 \sim 0.1$) and the other blackened with carbon ($\varepsilon_2 \sim 0.9$) on the side facing the beam. This should be done while keeping the characteristics of the outer surface of the thermocouple the same, with Δ roughly constant. Then, the two measured values would be

$$T'_1 = T + \Delta T_1 + \Delta$$
$$= T + \varepsilon_1 F + \Delta$$
$$\text{and} \quad T'_2 = T + \varepsilon_2 F + \Delta.$$

Therefore,

$$F = \left(\frac{T'_2 - T'_1}{\varepsilon_2 - \varepsilon_1}\right) \quad (5.7)$$
$$= \frac{(T'_2 - T'_1)}{0.8}.$$

If the thermocouple with the least systematic error ($\varepsilon \sim 0.1$) is chosen for measurements on lamps of the same design as that tested, the systematic error would be, using equation 5.7,

$$\Delta T_1 = \varepsilon_1 F \qquad (5.8)$$
$$= \frac{1}{8}(T_2' - T_1').$$

The value of ΔT_1 is rough, however, because values of ε_1 and ε_2 are uncertain. The effect on ΔT_1 of errors or changes in ε_1 and ε_2, $\delta\varepsilon_1$ and $\delta\varepsilon_2$, is calculated as follows.

$$\Delta T_1 = \varepsilon_1 F$$
$$= \varepsilon_1 \left(\frac{T_2 - T_1}{\varepsilon_2 - \varepsilon_1} \right) \qquad (5.9)$$

and by differentiating equation 5.9,

$$\frac{\delta(\Delta T_1)}{\Delta T_1} = \frac{\varepsilon_2}{\varepsilon_2 - \varepsilon_1}\left(\frac{\delta\varepsilon_1}{\varepsilon_1}\right) - \frac{\varepsilon_2}{\varepsilon_2 - \varepsilon_1}\left(\frac{\delta\varepsilon_2}{\varepsilon_2}\right)$$
$$\approx \frac{9}{8}\left(\frac{\delta\varepsilon_1}{\varepsilon_1} - \frac{\delta\varepsilon_2}{\varepsilon_2}\right).$$

Since ε_1 is an order of magnitude smaller than ε_2, the relative error $\delta\varepsilon_1/\varepsilon_1$ is far larger than $\delta\varepsilon_2/\varepsilon_2$. Thus, the uncertainty in ε_1 dominates and the relative error in ΔT_1 is roughly that in ε_1. At low values of emissivity, such as $\varepsilon_1 \sim 0.1$, we can assume an uncertainty of about 50% ($\delta\varepsilon_1/\varepsilon_1 \sim 1/2$).

Consequently, in using the chosen thermocouple (assuming $\varepsilon_1 = 0.1$, say) and lamp design, surface measurements should be corrected by $-(T_2' - T_1')/8$, from equation (5.8), and the systematic error associated with absorbed beam energy is covered by the above uncertainty in ΔT, i.e., $(T_2' - T_1')/16$.

5.9 Statistical treatments of random error

In the methods of mathematical statistics it is usual to employ the concept of an infinite population, the set of values that would result from a random experiment if repeated for an infinite time. Thus, the random components in a set of measurements are regarded as a small sample or sub-set of the population.

Populations of data are distributed in value in a manner dependent on the source of data and are described mathematically by 'distribution functions'. For example, random errors are usually considered part of a population having a 'Gaussian' or 'Normal' distribution.

5.9. STATISTICAL METHODS

When measurements are affected by significant random error it is necessary to estimate the 'true' value: that value that would have resulted if no random error had been present. The true value is the population mean, μ, and in this sense it is μ that measurement seeks to estimate. Of course, the infinite population is unavailable and μ is unknown. For a set of n readings, the jth being x_j, the best estimate of μ is the arithmetic mean, $\bar{x} = \sum x_j/n$. But, $\bar{x}$ is an estimate and its proximity to μ is a matter of chance. One measure of this difference, the random error in $\bar{x}$, can be had from the **standard deviation**, s, which is a very efficient measure of the spread of data, since it makes optimum use of the information. It is calculated from the following expression.

$$s = \sqrt{\left(\frac{\sum_{j=1}^{n}(x_j - \bar{x})^2}{n-1}\right)}. \tag{5.10}$$

It turns out that a little less than 68% of the data lies within $\pm s$ of $\bar{x}$. In other words, s is a measure of the likely difference between x_j and μ. What we need, though, is the likely difference between the measured mean, $\bar{x}$, and μ, i.e., some measure of the scatter in $(\bar{x} - \mu)$, which is smaller than s. One such measure is the standard deviation of the mean $\bar{x}$, often loosely referred to as the **standard error**, and equal to $s/\sqrt{n}$. On the other hand, it is more usual to use s in a different way—to form confidence limits for μ. This is done by considering the variable $\sqrt{n}(\bar{x} - \mu)/s$, which has a Student's t-distribution with $n-1$ degrees of freedom. The width of this distribution, containing $p\%$ of the values, is $\pm t_p$. Thus, we can write

$$\frac{\sqrt{n}}{s}(\bar{x} - \mu) = \pm t_p$$

or

$$\mu = \bar{x} \pm \frac{t_p s}{\sqrt{n}}. \tag{5.11}$$

The two limits, given by this expression, are known as the $p\%$ **confidence limits**, which define an interval centred on $\bar{x}$ and having $p\%$ chance of containing the true value μ. Values of t_p are given in Table 5.2.

Another approach is to use the range, R, of the measurements, $x_1, x_2, \ldots, x_n$. The range is defined as the difference between the largest and smallest values of x_j and, like s, it is a measure of the spread of data and can be used to calculate the uncertainty. The distribution for the variable $(\bar{x} - \mu)/R$ has been studied and its percentiles are given in reference [124]. From these it follows that $p\%$ of the population lies in the range

$$-m_p < \frac{\bar{x} - \mu}{R} < +m_p$$

and, therefore, that the $p\%$ confidence interval for μ is

$$\mu = \bar{x} \pm m_p R. \qquad (5.12)$$

Values of m_p are given in Table 5.2.

Table 5.2: Values of t_p and m_p as a function of number of degrees of freedom $\nu_{eff} = n - 1$: given for three values of probability, p, expressed as a percentage.

Number of Degrees of freedom ν_{eff}	Values of t_p			Values of m_p		
	$p = 90$	$p = 95$	$p = 99$	$p = 90$	$p = 95$	$p = 99$
2	2.92	4.30	9.92	0.89	1.30	3.01
3	2.35	3.18	5.84	0.53	0.72	1.32
4	2.13	2.78	4.60	0.39	0.51	0.84
5	2.02	2.57	4.03	0.31	0.40	0.63
6	1.94	2.45	3.71	0.26	0.33	0.51
7	1.89	2.36	3.50	0.23	0.29	0.43
8	1.86	2.31	3.36	0.21	0.26	0.37
9	1.83	2.26	3.25	0.19	0.23	0.33
11	1.80	2.20	3.11	0.16	0.19	0.28
13	1.77	2.16	3.01	0.14	0.17	0.24
15	1.75	2.13	2.95	0.12	0.15	0.21
17	1.74	2.11	2.90	0.11	0.14	0.19
19	1.73	2.09	2.86	0.10	0.13	0.18
30		2.04				
50		2.01				
∞	1.64	1.96	2.58	0.00	0.00	0.00

Obviously, the uncertainty based on m_p, given by equation 5.12, is easier to calculate than that based on t_p, from equation 5.11. The only disadvantage in using the former is that, on average, $m_p R$ is a little larger than $t_p s/\sqrt{n}$, because R is less efficient than s in representing the spread of data. This is the result of R being based on less information. Nevertheless, in my view, the added complication in using the standard deviation is not usually warranted, especially in temperature measurement. This is because random error tends to be swamped by the effects of systematic errors (see example on page 167). Hence the method based on the range is recommended.

Chapter 6

Multi-Site Temperature Measurement

6.1 Introduction

Elsewhere in this handbook the operating principles of various temperature (and other) sensors are discussed as are the sources of error that affect their use. In general, the articles cover measurement at any **one** site. By contrast, in this chapter, I deal with the additional problems that arise in multi-site measurement—if there is a need to measure temperature at more than one site throughout a specified region, how are the measurements arranged and what do we do with all the data? To measure temperature at n sites is not just the problem of repeating a single measurement n times.

In an oven or furnace, temperature will differ from point to point in the gas throughout the work space and at any one site it will vary in time because of control action. To measure the distribution of temperature a number of thermocouples may be installed and (a) they may significantly change the conditions in the chamber and (b) they are likely to be affected more by the perceived temperatures of the walls than by that of the gas. Moreover, objects placed within the work space will experience different temperatures again. Their surfaces, if of low thermal conductivity (paint, oxide, etc), will fluctuate in temperature more rapidly than will that of their bulk, and if their (total) emissivities differ from that of the test thermocouples so also will their temperatures. These issues must be considered when making multi-site measurements, but they won't be examined here—they are dealt with in section 3.10, beginning on page 103. In this chapter, I deal only with those concerns peculiar to multi-site measurement.

For any given situation requiring multi-site measurements, the reasons why the parameter of interest is expected to vary from site to site must first be understood. Then, parameters that describe the variation may be defined and a method for their measurement developed. The chosen approach, however, will be very dependent on the case in point—the aim of the measurements—and is best illustrated by example.

I have chosen to analyse the case where a heated enclosure is tested to see whether it is suitable for the thermal conditioning of a product covered by a standard materials specification, as described in the next section. By **enclosure** I mean any temperature-controlled, contained region that is nominally uniform in temperature, such as an oven, furnace or liquid bath.

The specific constraints of this problem/case and the method developed may not apply to other multi-site situations, but the approach taken will serve as a useful guide.

6.2 Materials specifications

Most materials specifications include a thermal conditioning requirement expressed in terms of a temperature range or tolerance—for example, in specifications for the accelerated ageing of rubbers and the heat treatment of metals. The tolerance usually takes the form: "... heat treat at a temperature of 505 ± 5°C for ...", or more generally,

$$\text{specified range} = T_o \pm \Delta T . \tag{6.1}$$

To condition the chosen objects (load) they are placed within a temperature-controlled enclosure, whose temperature is set at a nominal T_o, and held at temperature for a specified period. The question is: how may the enclosure be tested to see whether it is capable of meeting the temperature specification and, if a suitable test is devised and carried out, how are the specified treatment conditions best achieved.

The problems begin when we try to interpret the specification, equation (6.1). Firstly, the temperature specification is often so non-specific as to be ambiguous. Does it refer to the (average) temperature of the load and does $\pm \Delta T$ refer to how well this value is known? In which case, is ΔT the standard deviation or does it represent confidence limits? To many people, the temperature uniformity of a furnace is presumed—they speak of the furnace temperature (singular), as if there was but one value and it applies everywhere throughout the furnace. Then, does $\pm \Delta T$ merely reflect how well the control temperature is set or calibrated? Some oven specifications are so worded that $\pm \Delta T$ could be taken as referring only to the control cycle.

Clearly, we must establish the intended meaning of the temperature specification—by looking at the role played by temperature in forming or enhancing that materials property for which the treatment is given. For example, consider the solution heat treatment of aluminium alloys [125]. Here, the aim is to heat the material to a suitable temperature and hold it there long enough to form a homogeneous solid solution of the alloying elements. During solution heat treatment of an Al-rich AlCu alloy, the temperature should nowhere exceed the eutectic point (528°C for Al, 33 wt.% Cu), because localised melting may occur in segregated, Cu-rich regions. On the other hand, temperature must everywhere exceed the temperature at which all the available Cu will dissolve, given time. These limits must be further constrained—if the temperature is too near the eutectic point a loss in strength and ductility will occur, and if the temperature is too low, excessive soaking times are needed, because of the lower rates of solution and diffusion. Clearly the temperature throughout a load must be constrained, but the choice of temperature limits involves compromise and is thus subjective. The perceived difficulty in achieving tight temperature limits, for commercially-available enclosures, must be balanced against a loss in strength and an increase in soaking time.

From considerations such as this, it follows that temperature specifications are intended to cover conditions within the load, but the specified limits are somewhat rubbery—they should not be regarded as sharp truncations. It cannot be said that heat treatment at a temperature 0.1°C above the upper limit is 'bad' (unacceptable) while that at 0.1°C below the limit is good. Furthermore, the specified limits are likely to include safety margins.

Nevertheless, let's express the general temperature specification, represented by equation (6.1), in strict terms, as it was probably intended.

At no time during the specified treatment period shall the temperature at any point within the load be outside the range $T_o \pm \Delta T$.

This interpretation may well be what was intended by the authors of temperature specifications. Unfortunately, it cannot be tested—even if it were true, it cannot be shown to be so. The reasons for this difficulty are discussed in the next section.

6.3 Interpretation of temperature specifications

The strict interpretation of a temperature specification, given above, raises a number of questions that need addressing. It is not practical to measure temperature at numerous sites throughout each load as it is being treated. In some cases the insertion of temperature probes would be destructive and in all cases it would be too time consuming. So, the enclosure needs to be tested using a dummy/test run. There are three questions that follow as a direct consequence:

Q.1 Does the presumption that measured test values also apply to future occasions involve a prediction of some kind or do we simply include an uncertainty component to allow for **drift** in the behaviour of the enclosure and/or its controller? The most practical approach would seem to be (a) maintain the quality of operation of the enclosure and its instrumentation by an appropriate maintenance program and (b) include a relatively small component of uncertainty to cover drift, when the question of uncertainties (see below) is resolved.

Q.2 The specification refers to load temperatures. So, should the test be done with the enclosure empty or with a representative **load** present? If the test is performed on an empty enclosure the results will be more characteristic of the enclosure itself and thus will more readily reveal changes in its performance—but can any conclusions derived from an empty test be applied to a load? On the other hand, if it is tested with a load, the results will apply only to that particular load configuration—they will depend on the type, number and positions of the components and on their surface properties.

Q.3 The specification refers to the infinite number of sites throughout each load, or throughout the work space if tested empty. The test must obviously be done at a **finite** number of sites and will give a smaller range of values—a range that will depend on the number and positions of the sites. How are the sites to be chosen and how is the reduced spread of data compensated for? Do we simply choose a number large enough to be considered infinite, say 50 or more, or do we choose a more practical number and increase the measured spread in data by a statistically determined factor?

Furthermore, there are other questions of a more general nature, given below.

Q.4 Temperature is not a unique property of any one site. If the work space of an enclosure was a vacuum, temperature would have no meaning until a probe was inserted. Then, the temperature would simply represent the thermal state of the probe and would be dictated by the perceived temperature of the walls and the emissivity of the probe. If gas is added, the probe temperature would then be influenced to some extent by the temperature of the gas, depending on the degree of convection (see section 3.10). In any case, the measured temperature is very much dependent on characteristics of the **sensor**, such as its thermal capacity, surface properties and shape. If there is any significant control action,

6.3. INTERPRETATION

its temperature will vary in time and the measured fluctuation will be directly related to its thermal capacity. So, temperature in an enclosure is as arbitrary as the choice of sensor! Should this arbitrary-ness be handled by including a suitable uncertainty component? Or, should test sensors be specified, in which case, how?

Q.5 In general, those properties of the load being modified by the treatment will not be linear in time or in temperature. Hence, the developed properties of two samples treated with the same mean temperature will differ if, for example, one had experienced a temperature **cycle** and the other had not. The effect of excursions above the mean temperature would not be compensated for by like excursions below. Nevertheless, over a relatively small temperature interval, say $\Delta T/2$, the changes may be sufficiently linear for the temperature cycle to be ignored in deciding compliance with a temperature specification. So, the question arises: should the measured temperature cycle be discounted? Notice also that temporal fluctuations observed with a test probe depend on the response time of the probe and will in general differ from that of any one component of a load being treated.

Q.6 Temperature specifications do not make allowance for the **uncertainty** of measurement, let alone the uncertainties mentioned above. Even if the measured temperature range for the enclosure is very small in comparison to $2\Delta T$, it is impossible with certainty (100% probability, as implied by the specification) to claim the actual temperature range is less than $2\Delta T$ (section 5.2). The probability distribution function (Section 5.9) for the overall measurement error is most likely truncated, but even so its width would exceed $2\Delta T$. How should the uncertainties be treated when deciding compliance with a specification and what probability level should be assumed?

The resolution of the above must involve numerous arbitrary decisions. Temperature constraints given in materials specifications cannot be unambiguously interpreted, and what is needed is a **defined interpretation** that adequately incorporates the above issues. This has been achieved in AS 2853, for example, discussed in the next section.

Applying results obtained for an empty enclosure

With the exception of **Q.1** above, the various concerns are bypassed once a defined interpretation is adopted. If, in common with most enclosure test methods, including AS 2853, testing of empty enclosures is allowed or preferred, how does the user make use of the 'empty' results when heat treating a significant load? Of course, the option of conducting a test on a typical load may still be chosen, but often this is impractical.

Results obtained with small, defined sensors in an otherwise empty enclosure would, of course, apply to a load consisting of numerous, similarly-sized objects scattered throughout the chamber, e.g., small specimens being conditioned in a laboratory oven. But what of larger objects? Clearly, the presence of one or more significant objects in the work space would change the temperature distribution, as measured at 'empty' sites. This is due in part to the changed convective patterns and in part to the presence of additional hot surfaces—measured temperatures are a strong function of radiation transfer to surrounding surfaces, even for temperatures as low as 100°C (section 3.10).

For similar reasons, the presence of a load would change the relationship between the temperature of the enclosure-indicator probe and that at any point within the work space. It would also affect the dynamic response of the control probe to conditions in the enclosure, and thus the resultant control cycle.

A single, large load or many small components placed in a heap or a single basket is likely to experience a smaller range of temperatures throughout the load than would be measured for the empty enclosure. On the other hand, its mean temperature is likely to be significantly different. In a study [126] of a single muffle furnace, measured empty and with a variety of different loads, the temperatures within each load had a different mean and a much reduced spatial variation, but the temperatures still fell within the range observed when empty. If this result can be generalised, it would seem then that if the range for the empty enclosure fell within the temperature specification, subsequent treatments on loads are likely to be satisfactory. Of course, this would not apply if the load affected the proper/designed functioning of the enclosure.

The simple solution is to conduct regular tests using the empty state, as a convenient means of checking the operation of the enclosure and with the likelihood that the data will also apply to a variety of (sensible) loads. Then, an occasional test on a load would check the applicability of the empty data, e.g., occasionally, thermocouples could be added when a critical load is being treated.

The smoothing effect of a load should not be taken too far—as was commented in reference [127]: the "false assumption that the conductivity of metal will make up for the shortcomings of a furnace contributes heavily to the scrap piles .."

6.4 Standard AS 2853

6.4.1 Development of AS 2853

Prior to 1986, when AS 2853 [128] was first printed, there were no Standard Methods covering the testing and/or interpretation of temperature specifications in general. Naturally, enclosures were being tested, but very few of the test methods involved any accommodation of all six questions raised above.

There are a variety of standards (ASTM, BS and AS, etc) that deal exclusively with laboratory ovens, some for specific tasks, such as drying and ageing, and others for general use, giving a set of nominal tolerances and temperatures. They each have their own nomenclature and have differing ways of determining and expressing temperature variations. Those that give a test method usually specify the number, type and location of sensors—for example, 9 thermocouples each loaded with 3/16 inch steel spheres with one placed near each corner of the work space and one at the centre. By not being part of a materials specification and having nominal tolerances specified with an accompanying, self-consistent test method most of the problems of interpretation don't occur.

Before developing a method for testing enclosures there are two hurdles to be overcome. Firstly, there is a need to produce an interpretation that accommodates the above six issues (pages 192, 193), and secondly, there is the question of what to do with the measurement uncertainty?

For the latter there are at least two options. One option is to increase the spread of measured temperatures by a factor based on an estimate of the uncertainty. Then, the limits set by the temperature specification are compared with this expanded range, rather than with the maximum and minimum values measured. Alternatively, by using a suitably-defined interpretation, temperature specifications may be deemed to include an allowance for uncertainty. For example, an enclosure may be judged as suitable if the temperature limits of the specification are not exceeded by the extremes measured, using a defined set of sensors located in a specified way, provided the measurement uncertainty is less than some specified limit.

In the 60's the CSIRO National Measurement Laboratory (NML, then the National Standards Laboratory) was involved in testing enclosures. Having neither the mandate nor the authority to define how temperature specifications in general should be interpreted, the Laboratory chose to play safe and allow for uncertainty by expanding the measured range to give maximum and minimum likely values. To devise such a factor, a statistical study [126] was conducted and the effects of choosing and of using a finite number of test sites were assessed. One interesting observation is that for the same number of test sites, say 10 in a small oven, the measured spread of temperatures would be relatively small if the sites were chosen at random, it would be a little larger on average if a geometric pattern were used and it would be significantly larger again if the extremes were sought. For example, the overall temperature variation measured throughout the work space when the sites are chosen with the extremes in mind is, on average, 60% larger than that measured with randomly selected sites. In other words the most efficient way of utilising a relatively small number of sites is to seek extremes.

Moreover, the common approach of placing one thermocouple in each corner and one at the centre tends to bias the results low—there are more sensors placed to find cold spots than hot.

When an enclosure test [129] was performed at NML the issues raised in Q.3 to Q.6 above were tackled via the following interpretation:

At no time during the specified treatment period shall the temperature at any point within the load have a chance greater than 10% of experiencing a temperature outside the range $T_\circ \pm \Delta T$.

The number of test sites chosen for a work space volume V, in m^3, was $(9 + 3.5V)$ and the measured spatial, temporal and overall variations were increased to account for uncertainty. Then, the most critical parameters, the maximum and minimum likely temperatures were used for deciding compliance. Drift in enclosure performance was dealt with as indicated in item Q.1 above and the empty/loaded dilemma, Q.2, was treated using the arguments beginning on page 193.

Six industrial enclosures, including furnaces and salt baths, were tested in the above study [126]. Only one failed to comply with the relevant specifications when the uncertainties were ignored, whereas four of the six failed when the uncertainty factors were used. So, one of the consequences of taking account of the above six issues and of working to a meaningful interpretation was the higher 'failure' rate—not very popular.

Mid-range versus average temperature

When testing an enclosure there are two primary considerations—is the enclosure capable of achieving compliance with the relevant specification and, if so, how may the enclosure be best set to achieve compliance in subsequent uses?

The most efficient solution, the one that would allow the most enclosures to 'pass', is

- consider the enclosure as acceptable if the measured temperature range

6.4. STANDARD AS 2853

does not exceed $2\Delta T$ and
- choose the mid-range of all the measured data as that temperature that best represents the thermal state of the enclosure.

The reason for the latter is that the operator will, during a subsequent treatment, change the representative temperature given for the test to a value equal to T_o, by arranging a corresponding change in the indicated temperature. Here, the 'indicated' temperature is that value displayed on the temperature indicator and/or controller associated with the enclosure.

This is the reason why NML chose to use mid-range temperatures rather than average. The mid-range temperature for the enclosure is half the sum of the upper and lower extremes found during a test and its use is consistent with the above-stated aim in selecting test sites, of finding the extremes with equal emphasis.

Many enclosure-testing laboratories were using the average and I understand that their argument went as follows. Since the central value of a specification, T_o, is usually the temperature at which optimum change occurs, the conditions in an enclosure should be arranged so that the average temperature of the load is T_o or, maybe, that most of the load is at T_o. This approach would be fine if the temperature distribution was symmetric or that the overall range found was significantly smaller than $2\Delta T$. But they aren't. As shown below, if the average temperature of a load is adjusted to equal T_o part of the load may well be given an inadequate treatment even if the enclosure is capable of properly treating all the load. It is better to forgo the luxury of seeking optimum treatment for some of the load and instead seek to maximise the chance that all the load be given an acceptable treatment, at least. To illustrate this point consider the following case.

- The temperature specification is $500 \pm 5\,°C$.

- The indicated temperature was constant at $500\,°C$ during the test and within the enclosure the temperatures varied from 498 and $507\,°C$.

- Since the observed range, $9\,°C$, is less than that allowed, the enclosure is capable of achieving compliance.

- The mid-range temperature for the test is $(498 + 507)/2 = 502.5\,°C$, i.e., $2.5\,°C$ more than indicated. Thus, for subsequent treatments the operator would set the controller $2.5\,°C$ lower to give $497.5\,°C$ on the indicator and $500\,°C$ for the enclosure temperature. Clearly, all temperatures would be $2.5\,°C$ lower than before—495.5 to $504.5\,°C$ and within the specified range. Indeed, as expected, they are equispaced from the specified limits.

- During the test the average work space temperature was $504\,°C$, and if this was the value supplied as the enclosure temperature, the operator would set future treatments 4°C lower. Temperatures within the work space would then lie in the range 494 to 503°C and part of the load would be poorly treated.

An attempted solution

One problem with enclosure testing is the variety of test methods in use and thus the different outcomes that are possible for the same enclosure. Another, is that insufficient consideration had been given to the six issues raised above, and that when a method incorporating them was developed the chance of 'failing' the enclosure increased. An attempt to resolve this unsatisfactory position, at least for the Australian scene, began in April 1974 when the National Association of Testing Authorities (NATA) brought numerous interested parties together at the National Science Centre, Melbourne. The meeting formed a NATA-sponsored working group directed to develop a uniform approach to the question of testing enclosures. It was to formulate suitable nomenclature and produce a general test method.

In mid 1980 the resultant document was sent to Standards Association of Australia for consideration, and was the basis of their standard AS 2853 [128]. The rationale behind the method developed by the NATA working group is given below and the significant changes introduced in AS 2853 are given in the next section.

The NATA method hinges on two key elements.

1. The two extremes, T_{max} and T_{min}, within which lie all the measured data found in the work space during the test period. From these values we have the Overall Variation, $R = (T_{max} - T_{min})$ and the Measured Enclosure Temperature, MET $= (T_{max} + T_{min})/2$. The value of R is used to decide on compliance and MET is used by the operator to achieve optimum conditions in subsequent use.

2. A practical interpretation of temperature specification was defined as follows (except that x°C was used instead of ΔT).

   The allowed tolerance $\pm \Delta T$ is treated as equivalent to a 'specified overall variation' (R_o) in temperature of $2\Delta T$ and the enclosure is considered satisfactory if the measured overall variation does not exceed $2\Delta T$. Test measurements shall be made using sensors of defined response and accuracy placed at sites of specified number and location.

Built into the interpretation are the sensors, their number, locations and accuracy. These features will now be described.

6.4. STANDARD AS 2853

Sensors Since the temperature and temporal variations measured at any one site depend on the thermal capacity and emissivity of the sensor, a test sensor was defined as a thermocouple attached to a 5 mm long cylinder of metal of 5 mm diameter, with the metal and surface finish specified, depending on temperature. Alternatively, when such a sensor is impractical, a more appropriate choice may be made provided the sensor is described when reporting the results. For example, when the required accuracy is not achievable with thermocouples, resistance thermometry may be chosen.

Number of sites The number of sites shall be within 20% of a specified value, depending on the work space volume and how critical the treatment—small values of R require a larger number.

Site locations The aim in choosing sites was to find the extremes with equal emphasis. Thus, one site was at the centre of the work space and of the remainder, roughly half were to be located in positions expected to give the lowest values of temperature and the others where the highest values were expected. Although, as a precaution, the sites should also be distributed as widely as possible.

Uncertainties The uncertainties of measurement, rounding errors and, more importantly, the stochastic fluctuations arising from choice of sites would produce variations in test outcomes, if the test is repeated by the same or different laboratories. It follows that the overall uncertainty would contribute to passing unsatisfactory enclosures and failing satisfactory ones. These effects were minimised by limiting, for example, the total systematic uncertainty of measurement to $0.1R_\text{o}$ and the readability and rounding to $0.02R_\text{o}$. For the same reason, limits were placed on the repeatability of the enclosure indicator. As may be seen from Table 6.1, such limits are readily achievable for most enclosures. Conditioning treatments needing tighter control than $\pm 1\%$ in T usually require liquid baths and may be tested by resistance thermometers or differential thermocouples.

The majority view of the working group was that enclosures should be graded on performance, as judged from the overall variation. It was envisaged that future materials specifications could well refer to a required grade of treatment, based on R_o, which in turn is equivalent to the alternative tolerance ($\pm \Delta T$, with $R_\text{o} = 2\Delta T$), and that users may then have enclosures tested relative to the specified grade, rather than to R_o. So, a performance factor was devised:

$$f = \frac{100R}{100 + D}, \tag{6.2}$$

where D is the difference between the temperatures of the enclosure and ambient. The formula reflects two observations, that (1) R tends to be proportional to D—it is just as difficult to achieve ±5 at 500°C as ±10 at 1000°C; and that (2) near ambient, R tends not to depend much on temperature.

The grading scale was arbitrary and values of grade, G, increased by unity for each decrease in f by a factor of $\sqrt{2}$.

People responsible for technical tasks, such as achieving a critical change in a material or making measurements with electronic equipment, should be aware of the technical requirements of such actions. They should understand the meaning and significance of overall variation, electrical zero, gain and stability, etc. In my view, they would not see their equipment merely as items having arbitrary grade values. Consequently, pivotal to the NATA method is the specified overall variation R_o to be met by the enclosure. If a grade is specified instead, the equivalent value of R_o is first calculated and the various test constraints based on R_o are determined as before.

Table 6.1: Upper limits imposed on the (test) measurement uncertainties by the NATA method, for various enclosure tolerances ($\pm \Delta T/T$). Included is an example breakdown for components from the test thermocouples (TC) and the associated measuring instrument, with its resolution (0.02 R_o & proportional to temperature) given for type K at 100°C.

Temperature Tolerance (%)	Measurement Uncertainty (%)	Example Uncertainty Breakdown:		
		TC (%)	Instrument Uncertainty (%)	Resolution (μV)
±4	0.8	0.75[a]	0.28	5
±2	0.4	0.38[b]	0.14	2
±1	0.2	0.15[c]	0.13	1
±0.5	needs special consideration			

[a] corresponds to manufacturer's tolerance for standard-grade thermocouple wire.
[b] corresponds to manufacturer's tolerance for premium-grade thermocouple wire.
[c] best possible calibration uncertainty for types K and N thermocouples (see page 141).

NOTE: in practice, the above limits are readily achievable, since very few enclosures are governed by a tolerance of less than ±1% in temperature.

6.4. STANDARD AS 2853

6.4.2 The AS 2853 approach

The method of Standard AS 2853 [128], first printed in 1986, differs from that of the NATA working group (page 198) in four areas.

1. AS 2853 contains a more elaborate means of determining the number and positions of test sites.

2. It contains more options in choosing the test sensors—three categories of sensor are described, depending on the intended use for the enclosure. However, for each of the three categories, the standard allows the alternative option of using any other sensor provided it can be "demonstrated to have an equivalent or faster response time". In other words, assuming that 'faster' is to be read as 'smaller', a testing laboratory may, for convenience, standardise on the use of low tip-mass thermocouples that satisfy the minimum response-time requirement (Category 1). There is then the possibility of detecting rapid fluctuations in temperature that are not experienced by the intended loads and, thus, of inappropriately failing the enclosure!

3. It increased by a factor of 3 the maximum allowable uncertainty in the measurements and set a corresponding increase in the allowable resolution and in the rounding of results. Probably the limits were increased because of perceived difficulties in achieving the NATA values. If uncertainty levels were chosen as high as those allowed in AS 2853 the chance of an incorrect decision on compliance is quite high—the overall measurement uncertainty is about 50% of the critical parameter, R! It is better to keep to the NATA limits (they comply with AS 2853, as it sets upper limits) and use Table 6.1 as a guide.

4. The concept of grading the enclosure is given more weight—indeed, it is the pivotal parameter. One consequence is that for those cases (most) where a grade is not specified—the more technically relevant parameter, the overall variation, being specified instead—an equivalent grade must first be calculated before various test conditions can be determined (see below). Generally, the maximum allowed overall variation corresponding to the 'equivalent grade' is not equal to the specified overall variation! Fortunately, a grading statement need not be included in a report unless required by the user.

6.4.3 A sample enclosure test

As an example in the use of AS 2853, consider a laboratory oven that is to be used for treating material covered by the temperature specification $100 \pm 2.5\,°C$, i.e.,

$$\text{the specified overall variation, } R_o = 5\,°C.$$

- The first step is to find the equivalent grade (clause 9.1 of AS 2853), and this involves the following steps.
  (a) Assume a value for the ambient temperature for the coming test, say $20\,°C$. Thus, $D = 100 - 20 = 80\,°C$.
  (b) From equation (6.2), the equivalent 'grading factor' is $f = 2.78$.
  (c) The value of the 'maximum permissible grading factor' nearest to and larger than f is chosen (Table 1 in AS 2853). It is $f_m = 2.83$.
  (d) The grade equivalent to f_m is $G = 5$.

- The number of 'standard sites' is then obtained (clause 5.3.1 in AS 2853)

$$N = 3 + 3G^{0.6}V^{0.2} \quad \pm 10\,\%\,. \tag{6.3}$$

  Assuming the oven has internal dimensions of 0.5, 0.5 and 0.8 m, i.e., a volume of $V = 0.20\,\text{m}^3$, we have $N = 7.8$ to 9.6, or in practice

$$N = 8 \text{ or } 9.$$

- The maximum allowed systematic uncertainty in the measurement of each sensor temperature is $0.3R_o = 1.5\,°C$. As mentioned above, it is better to set the uncertainty limit even lower, say $0.5\,°C$. The calibration of test sensors is discussed on page 143.

- Options for the standard-sensor design (thermocouple wire diameter, tip mass, etc) are given in clause 5.2.1 of AS 2853.

- According to clause 5.3.2 (AS 2853) half the standard sites (4 or 5, depending on N, above) shall be located in a geometric pattern (Figure 2 of AS 2853) and of the remainder, half (2) shall be located where the lowest temperatures are to be expected and half for the highest.

- AS 2853 deals also with the measurement of other parameters and specifies the content of any measurement report, etc. As these issues are outside the scope of the chapter they are not discussed here.

At this stage the testing laboratory must consider the choice, the calibration and the use of test thermocouples (or other suitable sensors). Indeed, the laboratory must address the question of whether the probes are to be

6.4. STANDARD AS 2853

repeatedly re-used in other ovens?—it is a practical option. Such concerns are dealt with in the next section.

Assume that 9 sites were examined and that the results taken during a suitable test interval are those of Table 6.2. Here, the maximum and minimum temperatures recorded for each thermocouple are the extremes found at that site (during the test interval). Their difference (column 4) is the range of the variation seen at the site and the 'temporal variation' for the enclosure is defined as the maximum range observed, i.e., that for thermocouple 8. On the other hand, the 'spatial variation' is the range of the mid-range temperatures (column 5) for the nine sites. Finally, the most important parameter, the 'overall variation', the difference between the maximum and minimum measured temperatures (underlined in columns 2 and 3) is

$$R = 3.5°C,$$

and as R is less than R_o (5°C), the enclosure satisfies the specified tolerance. It also complies with the requirements of grade 5.

As to whether the enclosure is suitable for a higher grade classification, we need to work back from R and see if sufficient sites were examined and the uncertainty level used was low enough for the better performance category. From Table 6.2, $D = 99.2 - 21.4 = 77.8°C$ and from equation (6.2) the value of f equivalent to $R = 3.5$ is $f = 1.97$. This is less than the maximum allowed for grade 6 ($f_m = 2.00$ from Table 1 of AS 2853) and from equation (6.3) the number of sites required to be tested for this grade is 9 or 10. Since 9 sites were used and the measurement uncertainty was well below that required, the Standard allows the test report to indicate compliance at this better level as well.

This example highlights the advantages of using the largest value of N allowed by the Standard and a measurement uncertainty significantly less than the allowed limit. In the event that R exceeds R_o because of a local hot or cold spot, another benefit arises in having a large N. It is then possible, with the client's approval and some caution (see clauses 6.1 and 6.3 of AS 2853), to re-define the working space, make the resultant change in R and report a satisfactory outcome.

Choice and use of test thermocouples

The temperature distribution within a furnace or oven is usually measured with base-metal thermocouples and the procedure outlined in section 4.9 (page 143), for their calibration and use, may be used for any test temperature(s). However, if the enclosure is limited in use to temperatures below about 450°C, e.g., laboratory ovens, there are two other options.

Table 6.2: Data obtained in an oven tested for compliance with a temperature tolerance of ±2.5°C ($R_o = 5$). For each thermocouple position are shown the maximum and minimum values of measured temperature found during the test interval and their difference and mean (midrange of all data found at that position).

Thermocouple Position	Maximum (°C)	Minimum (°C)	Difference (°C)	Mid Range (°C)
1	<u>100.9</u>	100.8	0.1	<u>100.8</u>
2	100.5	100.3	0.2	100.4
3	99.9	99.8	0.1	99.8
4	99.0	98.8	0.2	98.9
5	100.2	100.0	0.2	100.1
6	100.5	100.2	0.3	100.4
7	100.2	100.0	0.2	100.1
8	97.9	<u>97.4</u>	<u>0.5</u>	<u>97.6</u>
9	100.8	100.7	0.1	100.8
Indicated Temperature	100.6	100.5	0.1	100.6
Ambient Temperature	21.6	21.2		21.4

Indicated Enclosure Temperature = 100.6°C
Measured Enclosure Temperature = 99.2°C

Spatial Variation = 3.2°C
Temporal Variation = 0.5°C
Overall Variation = 3.5°C

Ambient Temperature = 21.4°C

1. The thermocouples (new, used and/or pre-annealed) may be installed in the work space of the enclosure with the emf-producing zone of each (under door seal, for example) secured with adhesive tape, so that the zones don't move on re-opening the door. After conducting the enclosure test, *per se*, the thermocouples may then be calibrated, *in-situ*, by opening the door, placing all the thermocouple tips together, without disturbing (the positioning of) the emf-producing zones, and wrapping the bunched tips to the bulb of a calibrated resistance thermometer. The door is then closed and the enclosure temperature is taken through the sequence of values (approx.) used for the enclosure test, and held at each long enough to allow an intercomparison between the thermocouples and the resistance thermometer.

6.4. STANDARD AS 2853

2. Nickel-based thermocouples, type K or N, may be stabilised for use below about 450°C by a ~16 h anneal at 450°C (or longer, for greater stability—see page 47). If possible, the anneal should be given to the entire length of each thermocouple—the flexible thermocouples may be loosely coiled (diameter more than 300 mm), placed completely within the uniform-temperature zone of an oven/furnace and annealed at 450°C. The thermocouples would then be as homogeneous as they were initially and relatively stable, provided care is taken to avoid the effects of binders (etc.) mentioned on page 178. Alternatively, the thermocouples may be inserted in a suitable annealing furnace with sufficient immersion to uniformly anneal all sections likely to produce significant emf during the intended enclosure test(s). As a consequence of such a stabilising anneal, the thermocouples may be re-used in a variety of ovens—their calibrations may be done initially (but after the anneal), as described in section 4.9, or after use in an oven test, as described for option 1, above. Obviously, the inhomogeneity will worsen slightly with repeated use and, depending on the immersions and thermal histories, the calibrations will become more uncertain (depending on the period of anneal at 450°C). However, when in doubt, or if the enclosure test requires the minimum of uncertainty, the above-mentioned intercomparison with a resistance thermometer may be done.

Appendix A

The Thermodynamics of Thermoelectricity

A.1 General equations of flow

Thermodynamic treatments [130, 131] of thermoelectricity are based on Onsager's theorem. It states that if the generalised fluxes $\mathbf{J}_j$ are under the influence of generalised forces $\mathbf{X}_j$ then each $\mathbf{J}_j$ is linearly related to all $\mathbf{X}_j$, as follows
$$\mathbf{J}_j = \sum_i L_{ij} \mathbf{X}_j.$$

Furthermore, if $\mathbf{J}_j$ and $\mathbf{X}_j$ are so defined that the rate of irreversible entropy production per unit volume is

$$_i\dot{S}_v = \sum_i \mathbf{X}_i \cdot \mathbf{J}_i \tag{A.1}$$

then

$$L_{ij} = L_{ji}, \tag{A.2}$$

which is known as Onsager's reciprocity relation.

It can be shown that for a system with the two generalised fluxes, $\mathbf{J}$ (electrical) and $\mathbf{U}$ (thermal), the generalised forces are $\mathbf{E}/T$ and $\nabla(1/T)$, where $\mathbf{E}$ is the electric field. Hence

$$\begin{aligned}\mathbf{J} &= L_{11}\left(\frac{\mathbf{E}}{T}\right) + L_{12}\nabla\left(\frac{1}{T}\right) \\ \mathbf{U} &= L_{21}\left(\frac{\mathbf{E}}{T}\right) + L_{22}\nabla\left(\frac{1}{T}\right).\end{aligned}$$

The coefficients L_{ij} can be expressed in terms of the electrical and thermal conductivities, σ and κ, and the Seebeck and Peltier coefficients[1], S and Π. This is done by setting constraints for the above equations, so reducing them to known empirical relationships that define these parameters. It is instructive to do this in reverse, so let's consider Onsager's relations with the L_{ij} replaced, using $\nabla(1/T) = -\nabla T/T^2$ and some rearrangement:

$$\mathbf{J} = \sigma \mathbf{E} - \sigma S \nabla T \quad (A.3)$$
$$\mathbf{U} = \Pi \mathbf{J} - \kappa \nabla T. \quad (A.4)$$

To see how these equations reduce to known empirical 'laws' let's examine the following special cases:

1. with $\nabla T = 0$ in equation (A.3) we have Ohm's law

$$\mathbf{J} = \sigma \mathbf{E}, \quad (A.5)$$

2. with $\mathbf{J} = 0$ but $\nabla T \neq 0$ we have, from equation (A.3), the Seebeck effect

$$\mathbf{E} = S \nabla T, \quad (A.6)$$

3. with $\mathbf{J} = 0$ in equation (A.4) we have Fourier's law

$$\mathbf{U} = -\kappa \nabla T \quad (A.7)$$

4. and with $\nabla T = 0$ in equation (A.4)

$$\mathbf{U} = \Pi \mathbf{J}. \quad (A.8)$$

From the last expression, as $\mathbf{J}$ passes through a junction between metals A and B the difference in their Peltier coefficients leads to a discontinuity in heat flux,

$$\Delta \mathbf{U} = (\Pi_A - \Pi_B)\mathbf{J},$$

which appears at the junction. This is the Peltier effect.

A.2 The Kelvin relations

Thomson (Lord Kelvin) deduced the following relationships, and in so doing proposed a new effect, the Thomson effect, characterised by the coefficient μ,

$$\mu = T\frac{dS}{dT} \quad (A.9)$$
$$\Pi = TS. \quad (A.10)$$

---

[1] see notes on page 216 for comment on choice of terms and symbols.

A.3. SEEBECK EFFECT

The two expressions, relating the three thermoelectric coefficients[2] to each other and to the absolute temperature, T (in kelvin), are known as the Kelvin relations. Their derivation by Thomson was not rigorous, being based on inadequate theory, but they are correct. Their proof is not difficult, if the more recent techniques of irreversible thermodynamics are used. The first follows from the empirical definition of the Thomson coefficient, μ (given in section A.4). It is also a consequence of the law of conservation of energy [131]. The second is a consequence of Onsager's reciprocal relation, $L_{ij} = L_{ji}$, while evaluating the coefficients L_{ij} in terms of the empirically-defined parameters of equations (A.3) and (A.4).

The Peltier and Seebeck effects are always observed as the difference between the properties of two metals. Not so the Thomson effect—it occurs within a conductor and not necessarily near a junction. So the Thomson effect is especially valuable. It gives us the means of measuring directly the other thermoelectric properties of a metal, possibly its only practical use. Once μ has been determined as a function of the temperature T the Seebeck coefficient for the metal can be calculated from equation (A.9),

$$S(T) = \int_0^T \frac{\mu(T)}{T} dT \qquad (A.11)$$

and then the Peltier coefficient, $\Pi(T)$, follows from equation A.10. The use of expression (A.11) is discussed further in section A.5.

A.3 Seebeck effect

Onsager's relations give us some insight into the Seebeck effect. Consider an open-circuited wire, i.e., a wire in which no steady-state electric current can flow. If a temperature gradient, ∇T, develops, initially with $\mathbf{E} = 0$, we have from equation (A.3),

$$\mathbf{J} = -\sigma S \nabla T.$$

This flux represents a migration of charge, the flow becoming zero at the ends, or wherever $\nabla T = 0$, where charge accumulates. The reservoirs of charge then repel further migration, i.e., an electric field builds up and finally, when $\mathbf{J} = 0$, from equation (A.3),

$$\mathbf{E} = S \nabla T,$$

the expression for the Seebeck effect.

In the above treatment the Seebeck effect falls neatly into place and its relationship with the other electrical and thermal processes is clear. Phenomenologically, the Seebeck coefficient is seen as an intrinsic property similar

---
[2] see note 1 on page 216 for a comment on choice of symbols.

to, for example, the electrical and thermal conductivities. This approach offers no proof that the Seebeck effect, and thus the resulting Seebeck emf, develops within the conductor. We have effectively assumed that this is so by defining S as we did, as the proportionality constant between $\mathbf{E}$ and ∇T. In practice, the Seebeck effect is seen only when different materials are connected together, i.e., with junctions present; and the fact that it can be neatly and fully described thermodynamically using an intrinsic bulk property S is no proof of its existence. To seek the source of emf a microscopic view is required.

The mechanism of thermo-emf is described in the general theories of conduction for metals, where its relationship with contact potentials and the Fermi surface is clear. The presence of thermo-emf within the bulk of a material arises naturally when considering the behaviour of conduction electrons. On the other hand, no physical mechanism is available to support the notion of emf at junctions. The Seebeck coefficient [132, 133] appears as the rate of change in the Fermi level with temperature, measured about some arbitrary but constant zero, and is clearly an intrinsic property of a conductor. Contact potentials don't influence the operation of a thermocouple, but rather the reverse happens. When the temperature is uniform the contact potential difference between two conductors is the difference between their work functions. When the temperature is not uniform the contact potential difference is larger by the Seebeck emf.

A.4 Peltier and Thomson effects

Thermodynamic treatments of the Peltier and Thomson effects involve the calculation of changes in entropy. For example, equation (A.1) gives the rate of increase in entropy per unit volume for the ith irreversible process. Obert [131] employs the concept of entropy flow,

$$\mathbf{J}_s = \frac{\mathbf{U}}{T}.$$

The rate of entropy increase per unit volume is the rate of created entropy, the irreversible quantity $_i\dot{S}_v$ of equation (A.1), together with the rate at which the entropy flow decreases as it passes through the volume,

$$\dot{S}_v = {_i\dot{S}_v} - \nabla \cdot \mathbf{J}_s.$$

For example, the contribution to the last term from the electric flow $\mathbf{J}$ is from equation (A.4), with $\nabla T = 0$,

$$\nabla \cdot \mathbf{J}_s = \nabla \cdot \left(\frac{\Pi \mathbf{J}}{T}\right).$$

A.5. ABSOLUTE SEEBECK COEFFICIENTS

This contribution to the rate of entropy increase in a conductor is a function of its Peltier coefficient, Π, and the electric flux, **J**.

The above expression can be regarded as a change in the entropy (Peltier) flow vector, $\frac{\Pi \mathbf{J}}{T}$, and the Peltier effect at a junction or discontinuity in the conductor can then be considered as a change in entropy flow, $\delta(\frac{\Pi \mathbf{J}}{T})$. This led to the simple concept of 'Peltier flow' introduced on page 4.

Also, from equations (A.9) and (A.10) and with $\nabla \cdot \mathbf{J} = 0$, we have

$$\begin{aligned} \nabla \cdot \mathbf{J}_s &= \mathbf{J} \cdot \nabla S \\ &= \frac{dS}{dT} \mathbf{J} \cdot \nabla T \\ &= \frac{1}{T} \mu \mathbf{J} \cdot \nabla T. \end{aligned}$$

This demonstrates that reversible heat is liberated in a conductor whenever **J** and ∇T are nonzero and that it is proportional to both **J** and ∇T. The effect is also proportional to the Thomson coefficient, μ, which is defined by this expression.

The above analysis is an effective proof of Kelvin's first relation, $\mu = T(dS/dT)$, if the truth of the second, $\Pi = TS$, can be assumed,

We can also describe the Thomson effect without recourse to an entropy flow vector. The rate at which heat is liberated per unit volume is the irreversible production of heat, $\mathbf{J} \cdot \mathbf{E}$, together with the divergence of the heat flow vector, i.e.,

$$\dot{Q}_v = \mathbf{J} \cdot \mathbf{E} - \nabla \cdot \mathbf{U}$$

and from equations (A.3), (A.4) and (A.10) with $\nabla \cdot \mathbf{J} = 0$, we have

$$\dot{Q}_v = \frac{\mathbf{J}^2}{\sigma} - T \mathbf{J} \cdot \nabla S + \kappa \nabla^2 T.$$

Since $\nabla S = (dS/dT) \nabla T$, and using equation (A.9), it follows that

$$\dot{Q}_v = \frac{\mathbf{J}^2}{\sigma} - \mu \mathbf{J} \cdot \nabla T + \kappa \nabla^2 T. \tag{A.12}$$

The first term on the right refers to Joule heating and the last is associated with thermal conduction. The reversible component, $\mu \mathbf{J} \cdot \nabla T$, obviously represents the Thomson effect.

A.5 Absolute Seebeck coefficients

It is a simple matter to measure the Seebeck coefficient of one conductor relative to another. The two are joined as a thermocouple, one end is maintained at 0°C and the developed emf, $V(T)$, at various tip temperatures, T, is measured. From these data the relative Seebeck coefficient is $dV(T)/dT$, the slope of the curve $V(T)$ versus T.

But what of the absolute Seebeck coefficient of any one metal? It cannot be measured directly, although, if the metal and a superconductor ($S = 0$) are combined as a thermocouple the absolute coefficient of the metal would equal the coefficient for the combination. But such a method is limited to the temperature range in which superconductivity occurs. Instead, it is preferable to measure the Thomson coefficient, $\mu(T)$, for the metal of interest, a somewhat difficult exercise, and then calculate the absolute Seebeck coefficient using equation (A.11), restated here:

$$S(T) = \int_0^T \frac{\mu(T)}{T} dT,$$

where T is in kelvin (K).

This technique has been applied to several pure reference materials to form 'absolute scales of thermoelectricity', e.g., reference data on Pb, Pt and W cover the range from ~ 0 to at least 2400 K (~ 2100°C). It is then a simple step to obtain the absolute coefficients for other conductors, by first measuring the relative coefficient for each conductor against a suitable reference material. The most accurate reference data for Pb have been obtained for the temperature range ~ 0 to 550 K [134, 15], for Pt, over the range 273 to 1600 K [17] and for W from 273 to 1800 K [17]. Accurate reference data for Cu and Au are also available [15].

Platinum is the most commonly used reference material for the metals of thermocouples. Its absolute coefficient is given in Figure A.1 and Table A.1 for temperatures from 5 K to its freezing point, 1768°C. The data were taken from reference [135] for temperatures up to 80 K and reference [17] for 273 to 1600 K (1327°C). From 80 to 273 K graphical interpolation was used, with other data [136, 16] as a guide, and beyond 1600 K the curve was extrapolated with the help of less accurate data [16].

Values of emf and Seebeck coefficient, relative to Pt, are available [12, 13] for the common thermocouple metals and the individual coefficients for these metals can thus be determined using the absolute scale for Pt. Indeed, this was how the data on absolute Seebeck coefficients, given in section 2.3, were obtained.

Once the (absolute) Seebeck coefficient is known, values for the other thermoelectric coefficients, μ and Π, can be calculated from the Kelvin

A.6. SOME CALCULATIONS FOR COPPER

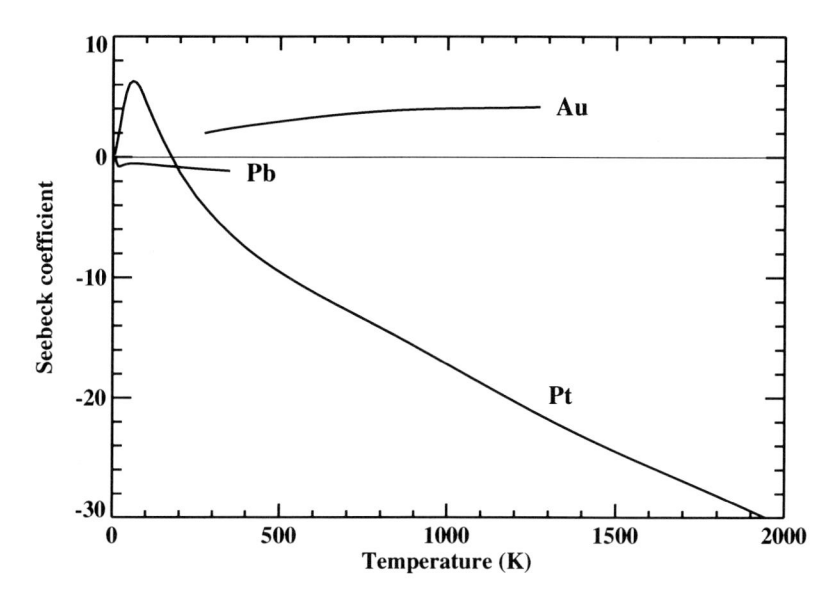

Figure A.1: The absolute Seebeck coefficient (μVK^{-1}) of platinum (Pt) as a function of temperature. Data for lead (Pb) and gold (Au) are included for comparison.

relations, discussed in section A.2,

$$\mu = T\frac{dS}{dT}$$
$$\Pi = TS.$$

For example, the values given in Table 1.1 on page 7, for a variety of metals at 20°C, were calculated in this way. The derivative dS/dT was obtained graphically.

A.6 Some calculations for copper

The relative magnitudes of some thermal, electric and thermoelectric parameters will be calculated for Cu at 20°C, as required in section 1.2. First we will discuss the conduction electrons using the free-electron model [137].

Only those valence electrons within $\sim kT$ (0.02 eV) of the Fermi energy (~ 7 eV) are available for conduction processes, i.e., only the fraction 0.02/7 or 1/350. Further, the relaxation time, τ, and the average speed, $\bar{v}$, of these

Table A.1: The absolute thermoelectric scale for platinum: values of Seebeck coefficient, S (μV K^{-1}), up to its melting point.

Temperature (K)	S	Temperature (°C)	S
5	0.2	0	−4.0
10	0.9	100	−6.9
20	2.4	200	−9.0
30	4.0	300	−10.8
40	5.3	400	−12.3
60	6.3	600	−15.2
80	5.8	800	−18.3
100	4.5	1000	−21.4
150	1.4	1200	−24.1
200	−1.3	1400	−26.6
250	−3.3	1600	−29.1
300	−4.9	1768	−31.1

electrons are

$$\tau = 2 \times 10^{-14} \text{ s}^{-1}$$
$$\bar{v} = 1.6 \times 10^8 \text{ cm s}^{-1}$$
$$= 5.8 \times 10^6 \text{ km h}^{-1}$$

and the number of collisions per second ($1/\tau$) is 5×10^{13} s^{-1}.

When an electric field, E, is applied, only those electrons able to accept an increase in energy are affected, about 0.3% of them (see above). The resultant drift velocity, for electrons of mass m and charge e, is

$$v_D = e\tau \frac{E}{m}.$$

The electron mobility (v_D/E and thus $e\tau/m$) has the value ~ 30 cm^2 V^{-1}s^{-1} and from Ohm's law, equation (A.5), with $\sigma = 5.9 \times 10^5$ Ω^{-1} cm^{-1} for the conductivity, we have

$$v_D = \frac{e}{m}\tau\frac{J}{\sigma}$$
$$= 5.1 \times 10^{-5} J \text{ cm s}^{-1}. \qquad (A.13)$$

A.6.1 The electric case

Consider a loop of copper at uniform temperature ($\nabla T = 0$) having a length L and a cross-sectional area of A. If a DC electric current, i, flows in the loop

A.6. SOME CALCULATIONS FOR COPPER

there will be an accompanying heat flux, as seen by equation (A.8), equivalent to a heat current of Πi, and referred to here as the Peltier flow. Such motion by the electrons is also 'lossy' and the total Joule (resistive) heat irreversibly liberated in the loop is $i^2 L/\sigma A$. With $i = 10$ A, $L = 100$ cm, $A = \pi/400$ cm² (1 mm diam. wire) and, for copper, $\Pi = 0.56$ mW and $\sigma = 5.9 \times 10^5\ \Omega^{-1}\,\text{cm}^{-1}$, we have

$$\begin{aligned}\text{Joule heat} &= 2.15\ \text{W} \\ \text{Peltier flow} &= 5.6\ \text{mW.}\end{aligned}$$

Also, $J = i/A = 4000/\pi = 1273\,\text{A cm}^{-2}$ and from equation (A.13), $v_D = 6.5 \times 10^{-2}\,\text{cm s}^{-1}$ or $3.9\,\text{cm min}^{-1}$.

If a temperature gradient, dT/dx, of $100\,\text{K cm}^{-1}$ is now arranged somewhere in the loop there will be the additional, reversible heat transfer between the wire and its surroundings due to the Thomson effect. From equation (A.12), with dT/dx as the one-dimensional equivalent to ∇T, the Thomson heat per unit length is, with $\mu = 1.9 \times 10^{-6}$ V and $i = 10$ A,

$$\mu i \frac{dT}{dx} = 1.9\ \text{mW cm}^{-1}.$$

A.6.2 The thermal case

Consider a piece of copper with no applied electric field, i.e. with $E = 0$, initially at least. If a temperature gradient, dT/dx, instantly develops along the copper we can see what happens from the Onsager relations (A.3) and (A.4).

Initially, with $E = 0$, the electric current density in copper at 20°C ($\sigma = 5.9 \times 10^5\,\Omega^{-1}\,\text{cm}^{-1}$ and $S = 1.9 \times 10^{-6}\,\text{VK}^{-1}$) is

$$\begin{aligned} J &= \sigma S \frac{dT}{dx} \\ &= 1.1 \frac{dT}{dx}\ \text{A cm}^{-2}.\end{aligned}$$

The drift velocity, v_D, is given by equation (A.13) and the Peltier flux is ΠJ ($\Pi = 0.56$ mW),

$$\begin{aligned} v_D &= 5.4 \times 10^{-5} \frac{dT}{dx}\ \text{cm s}^{-1} \\ \Pi J &= 5.9 \times 10^{-4} \frac{dT}{dx}\ \text{W cm}^{-2}.\end{aligned}$$

The electric flow causes a charge imbalance and thus an electric field. The build up of charge continues until the field balances any tendency for electron drift, i.e. until $J = 0$. Thus the above values of J, v_D and ΠJ are initial values and they exponentially decay to zero as the electric field develops. The process takes an insignificantly small time and from equations (A.3) and (A.4) we have for the steady state, when $J = 0$,

$$E = S\frac{dT}{dx}$$
$$= 1.9 \times 10^{-6} \frac{dT}{dx} \text{ V cm}^{-1}$$

and

$$U = -\kappa \frac{dT}{dx}$$
$$= -4.0 \frac{dT}{dx} \text{ W cm}^{-2},$$

since for copper, $\kappa = 4.0 \text{ W cm}^{-1}\text{K}^{-1}$.

The initial Peltier flux is considerably smaller than the 'normal' heat flux U. Taking the above values for ΠJ and U, both being proportional to dT/dx, we have

$$\frac{\Pi J}{U} = -1.5 \times 10^{-4} \quad (\text{or } -0.015\%).$$

By way of summary, consider $dT/dx = 100 \text{ K cm}^{-1}$ and a 1 cm^{-1} cross section. If the temperature gradient could be applied instantaneously, an electric current of 110 A would appear requiring a drift velocity of $5.4 \times 10^{-3} \text{ cm s}^{-1}$ and an associated Peltier flow of 0.059 W. These flows exponentially decay and steady state conditions will be established almost instantaneously. Then the electric field will be $190 \,\mu\text{V cm}^{-1}$ and a heat flow of 400 W will be present along the rod.

Notes

1. the symbols S, Π and μ are commonly used for the Seebeck, Peltier and Thomson coefficients and yet they are not consistent with each other. To symbolise the three thermoelectric effects using the initials S, P and T of their discoverers would clash with the use of S for entropy and T for temperature. Alternatively, I could have chosen the Greek symbols σ, π and τ, but σ is commonly used for the electrical conductivity and τ for a characteristic time, such as the relaxation time. My choice arose from an inability to find a consistent set and a strong preference for S as the Seebeck coefficient.

2. I chose to use the term 'Seebeck coefficient' in preference to the commonly used alternatives, 'thermoelectric power' and 'thermopower'. Firstly, it is consistent with the terms Peltier coefficient and Thomson coefficient, used to quantify the other thermoelectric processes, as each is named after its discoverer. Secondly, I avoid the use of 'power', here meaning strength, but normally the rate of transfer of energy.

Appendix B

Standard Reference Functions for Thermocouple Emf

B.1 Introduction

As of 1 January 1990, values of temperature should be given in terms of ITS–90, the International Temperature Scale of 1990 [138, 139]. This differs qualitatively and in definition from the previous scale, the International Practical Temperature Scale of 1968 (IPTS–68). As a result, values of temperature for the same thermal event differ slightly from values on the earlier scale (by less than 0.2°C up to 1000°C, with the difference becoming increasingly negative at higher temperatures—reaching $T_{90} - T_{68} = -1.5$°C at 3000°C, for example [139]).

The change in temperature scale necessarily caused changes in the emf-temperature relationship for each thermocouple. The changes, indicated in Table B.1, are not significant for the base-metal thermocouple types, being the equivalent of ≤ 0.1°C up to 900°C. When expressed as a percentage, the change in emf is $\leq 0.025\%$ up to 1000°C, which is smaller than the level of inhomogeneity in the as-manufactured thermocouple. For the rare-metal thermocouples, types R, S and B, the changes in reference emf are only just significant, and then, only if they are used as calibrated standards.

B.2 Reference functions

ITS–90 based reference functions [21, 22] for the letter-designated thermocouple types have had international acceptance. The functions are, with one exception, polynomials:

$$V_{ref} = \sum_{j=0}^{n} c_j T^j, \qquad (B.1)$$

where the emf, V_{ref} in μV, is the reference emf for a thermocouple with its tip at temperature T, in °C, and having a cold junction at 0°C.

The exception is the expression representing type K thermocouples in the range 0 to 1372°C. It has, in addition to the polynomial, an exponential term:

$$V_{ref} = \sum_{j=0}^{n} c_j T^j + d_1 \exp\left\{-\frac{1}{2}\left(\frac{T-d_2}{65}\right)^2\right\}, \qquad (B.2)$$

where the defined values [21] of d_1 and d_2 are 118.5976 and 126.9686, respectively. This term has a peak contribution of $\sim 119\,\mu$V ($= d_1$) at ~ 127°C ($= d_2$) and little effect above 300°C. The coefficients may be rounded to $d_1 = 119$ and $d_2 = 127$ with little error in V_{ref} ($\leq 0.4\mu$V, equivalent to ≤ 0.01°C), although less rounding is suggested in Table B.4.

It is recommended that when evaluating equation (B.1), and the polynomial in (B.2), that each $c_j T^j$ be not separately calculated and summed. Instead, the operations should be nested, i.e., placed in the following form—it involves less computer time. Furthermore, double precision should be used.

$$V_{ref} = c_0 + (c_1 + \ldots (c_{n-3} + (c_{n-2} + (c_{n-1} + c_n T)T)T)\ldots)T \qquad (B.3)$$

This looks more palatable when re-expressed in programming form.

Step 1 : $E = 0$ and $j = n$
2 : $E \Leftarrow c_j + E$
3 : if $j = 0$ go to step 6
4 : $E \Leftarrow ET$ and $j \Leftarrow j - 1$
5 : go to step 2
6 : $V_{ref} = E$.

The standard reference functions are limited to particular temperature regions, so that two or three different polynomials are required to cover the full operating range of each thermocouple. Each coefficient, c_j, has 11 digits (12 for types R and S) and up to 20 coefficients are needed to fully describe the relationship between emf and temperature for any one thermocouple type. As with d_1 and d_2, roughly half the specified digits are not significant, i.e., their presence has an effect that is small when compared to the achievable calibration uncertainty of thermocouples.

Rounded coefficients for the standard reference functions to suit all letter-designated thermocouple types (B, E, J, K, N, R, S and T) are given in Tables B.3 to B.6. When using the rounded coefficients, the resultant reference equations are equally smooth representations of thermocouple behaviour as are the defined (above) reference functions, and differ from the defined set by less than that indicated in Table B.2. The difference is negligible. For example, for type R thermocouples, the use of the rounded coefficients causes in V_{ref} an error less than the equivalent of 0.01 °C.

The defined reference functions are also commonly represented by their tabulated equivalent: the standard reference tables (see section 1.4)—values of emf generated by these functions, usually given to the nearest 1 μV every 1 °C, e.g., in ASTM E230–93. For convenience, an abbreviated set of reference data is given in Tables B.7 to B.14—they contain the standard values in increments of 10 °C. Values of temperature (or emf) can be obtained for intermediate values of emf (or temperature) by linear interpolation, with little error.

For thermocouple types B, R and S, the interpolation error is less than ~0.5 μV for any 10 °C interval over the full temperature ranges of the above-mentioned tables. For the base-metal types, E, J, K, N and T, the error is < 1 μV when interpolating for any positive temperature, but below 0 °C the interpolation error increases with decreasing temperature, to about 3 μV at −200 °C, for example.

B.3 Converting V to T

The standard thermocouple reference functions are in the form $V = f(T)$, for converting temperature values into emf's. However, the thermocouple user more often wishes to operate in reverse—to convert a value of thermocouple emf into temperature. There are two options. Firstly, there is the use of 'inverse functions', for example, those given in reference [21], which are equivalent to the standard reference functions with an uncertainty better than ~ 0.05 °C. Coefficients for thermocouple types T and K are given in tables B.15 and B.16.

Secondly, an iterative process may be used to effectively apply the standard reference functions in reverse. For example, the procedure given below produces a value of T that is equivalent to a chosen value of emf, V_m, according to the reference function, equation (B.1). The procedure begins with a convenient approximation for expressing T in terms of V: the linear equation in step 1 below. Here, the constants A, B and S (a rough value to represent the average Seebeck coefficient) are chosen to suit the thermocouple type and the temperature range of interest. The chosen values are not critical—their main effect is on the speed of iteration. Step 3 sets the upper value for the error of iteration, in this case to about 0.01 °C.

$$
\begin{aligned}
\text{Step 1} \;&:\; T = A + (V - B)/S \\
2 \;&:\; \text{from } T, \text{ evaluate } V_{ref} \text{ using equation (B.1)} \\
3 \;&:\; \text{if } |V_{ref} - V| < S/100 \text{ go to step 6} \\
4 \;&:\; T \Leftarrow T - (V_{ref} - V)/S \\
5 \;&:\; \text{go to step 2} \\
6 \;&:\; \text{end.}
\end{aligned}
$$

REFERENCE FUNCTIONS

Table B.1: The difference between the value of emf given by the standard reference functions, developed from the 1990 data (equation B.1), and that from the 1968. The values, in μV, were determined at selected values of temperature (T) for the thermocouple types T, E, K, N, R and S.

Temperature (°C)	\multicolumn{6}{c}{$E_{90}(T) - E_{68}(T)$}
	T	E	K	N	R	S
0	0.0	0.0	0.0	0.0	0.0	0.0
100	1.2	1.8	0.9	0.5	0.2	0.6
200	2.0	2.7	1.9	1.5	0.5	0.8
300	2.1	3.1	1.2	1.2	0.5	0.4
400	2.8	3.3	1.7	1.5	0.5	−0.4
500		6.4	4.1	3.5	0.7	−0.3
* ∼600		8.7	4.5	4.1	1.7	1.8
800		−4.9	−1.4	−1.7	1.0	0.1
1000		15.3	6.9	7.4	2.8	2.4
1200			10.2	10.3	4.1	3.5
1400					5.4	4.7
1600					6.6	5.7

* Temperature near 600°C where the emf difference peaks— the peak values are given in this row of the table.

Table B.2: Maximum error (μV) arising from the use of rounded coefficients, given in Tables B.3 to B.6, in the reference equations (B.1) and (B.2).

Thermocouple Type	\multicolumn{3}{c}{Max. error for temperatures:}
	−200* to 200°C	200 to 1000°C	above 1000°C
B	0.01	0.1	0.2
R	0.01	0.1	0.1
S	0.01	0.1	0.1
E	0.1	0.5	—
J	0.1	0.2	—
K	0.1	0.6	0.9
N	0.1	0.4	0.4
T	0.1	0.1	—

* −50°C for types B, R and S.

Table B.3: Coefficients of reference equation (B.1), for converting temperature (ITS-90) to emf in μV. The coefficients have been rounded with negligible effect (see text).

Type B

Temperature Range	j	c_j	Temperature Range	j	c_j
0 to 630°C	0	0	630 to 1820°C	0	-3.8938×10^3
	1	-2.465×10^{-1}		1	2.85717×10^1
	2	5.904×10^{-3}		2	-8.48851×10^{-2}
	3	-1.326×10^{-6}		3	1.578528×10^{-4}
	4	1.567×10^{-9}		4	-1.683534×10^{-7}
	5	-1.694×10^{-12}		5	1.110979×10^{-10}
	6	6.299×10^{-16}		6	$-4.451543 \times 10^{-14}$
				7	9.89756×10^{-18}
				8	-9.3791×10^{-22}

Type E

Temperature Range	j	c_j	Temperature Range	j	c_j
−270 to 0°C	0	0	0 to 1000°C	0	0
	1	5.8666×10^1		1	5.8666×10^1
	2	4.541×10^{-2}		2	4.50323×10^{-2}
	3	-7.7998×10^{-4}		3	2.8908×10^{-5}
	4	-2.58002×10^{-5}		4	-3.3057×10^{-7}
	5	-5.94526×10^{-7}		5	6.50244×10^{-10}
	6	-9.321406×10^{-9}		6	-1.91975×10^{-13}
	7	$-1.02876055 \times 10^{-10}$		7	-1.25366×10^{-15}
	8	$-8.0370123 \times 10^{-13}$		8	2.148922×10^{-18}
	9	$-4.39794974 \times 10^{-15}$		9	$-1.438804 \times 10^{-21}$
	10	$-1.64147764 \times 10^{-17}$		10	3.59609×10^{-25}
	11	$-3.96736195 \times 10^{-20}$			
	12	$-5.5827328 \times 10^{-23}$			
	13	$-3.465784 \times 10^{-26}$			

REFERENCE FUNCTIONS

Table B.4: Coefficients of reference equation (B.1), or (B.2) if type K and above 0°C, for converting temperature (ITS–90) to emf in μV. The coefficients have been rounded with negligible effect (see text).

Type J

Temperature Range	j	c_j
−210 to 760°C	0	0
	1	5.0381×10^{1}
	2	3.0476×10^{-2}
	3	-8.5681×10^{-5}
	4	1.3228×10^{-7}
	5	-1.7053×10^{-10}
	6	2.09481×10^{-13}
	7	-1.2538×10^{-16}
	8	1.5632×10^{-20}

Type K

Temperature Range	j	c_j	Temperature Range	j	c_j
−270 to 0°C	0	0	0 to 1370°C	0	-1.76×10^{1}
	1	3.945×10^{1}		1	3.8921×10^{1}
	2	2.3622×10^{-2}		2	1.8559×10^{-2}
	3	-3.2859×10^{-4}		3	-9.9458×10^{-5}
	4	-4.9905×10^{-6}		4	3.18409×10^{-7}
	5	-6.7509×10^{-8}		5	$-5.607284 \times 10^{-10}$
	6	-5.74103×10^{-10}		6	5.607506×10^{-13}
	7	$-3.108887 \times 10^{-12}$		7	-3.20207×10^{-16}
	8	-1.04516×10^{-14}		8	9.71511×10^{-20}
	9	-1.98893×10^{-17}		9	-1.21047×10^{-23}
	10	-1.63227×10^{-20}		$d_1 = 118.6$	
				$d_2 = 127$	

Table B.5: Coefficients of reference equation (B.1), for converting temperature (ITS-90) to emf in μV. The coefficients have been rounded with negligible effect (see text).

Type N

Temperature Range	j	c_j	Temperature Range	j	c_j
-270 to $0\,°\!C$	0	0	0 to $1300\,°\!C$	0	0
	1	2.616×10^1		1	2.5929×10^1
	2	1.096×10^{-2}		2	1.571×10^{-2}
	3	-9.384×10^{-5}		3	4.3826×10^{-5}
	4	-4.641×10^{-8}		4	-2.52612×10^{-7}
	5	-2.63×10^{-9}		5	6.43118×10^{-10}
	6	-2.2653×10^{-11}		6	$-1.006347 \times 10^{-12}$
	7	-7.609×10^{-14}		7	9.974534×10^{-16}
	8	-9.342×10^{-17}		8	$-6.086324 \times 10^{-19}$
				9	2.084923×10^{-22}
				10	-3.06822×10^{-26}

Type R

Temperature Range	j	c_j	Temperature Range	j	c_j
-50 to $1064\,°\!C$	0	0	1064 to $1665\,°\!C$	0	2.9515×10^3
	1	5.2896×10^0		1	-2.5206×10^0
	2	1.3917×10^{-2}		2	1.59565×10^{-2}
	3	-2.3886×10^{-5}		3	-7.64086×10^{-6}
	4	3.5692×10^{-8}		4	2.0531×10^{-9}
	5	-4.62348×10^{-11}		5	-2.934×10^{-13}
	6	5.0078×10^{-14}			
	7	-3.7311×10^{-17}			
	8	1.57716×10^{-20}			
	9	-2.8104×10^{-24}			

REFERENCE FUNCTIONS

Table B.6: Coefficients of reference equation (B.1), for converting temperature (ITS-90) to emf in μV. The coefficients have been rounded with negligible effect (see text).

Type S

Temperature Range	j	c_j	Temperature Range	j	c_j
-50 to $1064\,°C$	0	0	1064 to $1665\,°C$	0	1.329×10^3
	1	5.4031×10^0		1	3.3451×10^0
	2	1.25934×10^{-2}		2	6.548×10^{-3}
	3	-2.32478×10^{-5}		3	-1.64856×10^{-6}
	4	3.2203×10^{-8}		4	1.3×10^{-11}
	5	-3.31465×10^{-11}			
	6	2.55744×10^{-14}			
	7	-1.25069×10^{-17}			
	8	2.7144×10^{-21}			

Type T

Temperature Range	j	c_j	Temperature Range	j	c_j
-270 to $0\,°C$	0	0	0 to $400\,°C$	0	0
	1	3.8748×10^1		1	3.8748×10^1
	2	4.4194×10^{-2}		2	3.3292×10^{-2}
	3	1.1844×10^{-4}		3	2.0618×10^{-4}
	4	2.0033×10^{-5}		4	-2.18822×10^{-6}
	5	9.0138×10^{-7}		5	1.09969×10^{-8}
	6	2.2651156×10^{-8}		6	-3.08158×10^{-11}
	7	$3.6071154 \times 10^{-10}$		7	4.54792×10^{-14}
	8	3.849394×10^{-12}		8	-2.75129×10^{-17}
	9	$2.8213522 \times 10^{-14}$			
	10	$1.42515947 \times 10^{-16}$			
	11	$4.87686623 \times 10^{-19}$			
	12	$1.07955393 \times 10^{-21}$			
	13	$1.3945027 \times 10^{-24}$			
	14	$7.9795154 \times 10^{-28}$			

Table B.7: Standard reference data for **type B** thermocouples. Data are values of thermocouple emf, in μV, for various tip temperatures, assuming a cold junction at 0°C.

°C	0	10	20	30	40	50	60	70	80	90	100
0	0	-2	-3	-2	0	2	6	11	17	25	33
100	33	43	53	65	78	92	107	123	141	159	178
200	178	199	220	243	267	291	317	344	372	401	431
300	431	462	494	527	561	596	632	669	707	746	787
400	787	828	870	913	957	1002	1048	1095	1143	1192	1242
500	1242	1293	1344	1397	1451	1505	1561	1617	1675	1733	1792
600	1792	1852	1913	1975	2037	2101	2165	2230	2296	2363	2431
700	2431	2499	2569	2639	2710	2782	2854	2928	3002	3078	3154
800	3154	3230	3308	3386	3466	3546	3626	3708	3790	3873	3957
900	3957	4041	4127	4213	4299	4387	4475	4564	4653	4743	4834
1000	4834	4926	5018	5111	5205	5299	5394	5489	5585	5682	5780
1100	5780	5878	5976	6075	6175	6276	6377	6478	6580	6683	6786
1200	6786	6890	6995	7100	7205	7311	7417	7524	7632	7740	7848
1300	7848	7957	8066	8176	8286	8397	8508	8620	8731	8844	8956
1400	8956	9069	9182	9296	9410	9524	9639	9753	9868	9984	10099
1500	10099	10215	10331	10447	10563	10679	10796	10913	11029	11146	11263
1600	11263	11380	11497	11614	11731	11848	11965	12082	12199	12316	12433
1700	12433	12549	12666	12782	12898	13014	13130	13246	13361	13476	13591
1800	13591	13706	13820								

REFERENCE FUNCTIONS

Table B.8: Standard reference data for **type E** thermocouples. Data are values of thermocouple emf, in μV, for various tip temperatures, assuming a cold junction at 0°C.

°C	0	10	20	30	40	50	60	70	80	90	100
−200	−8 825	−9 063	−9 274	−9 455	−9 604	−9 718	−9 797	−9 835			
−100	−5 237	−5 681	−6 107	−6 516	−6 907	−7 279	−7 632	−7 963	−8 273	−8 561	−8 825
0	0	−582	−1 152	−1 709	−2 255	−2 787	−3 306	−3 811	−4 302	−4 777	−5 237
0	0	591	1 192	1 801	2 420	3 048	3 685	4 330	4 985	5 648	6 319
100	6 319	6 998	7 685	8 379	9 081	9 789	10 503	11 224	11 951	12 684	13 421
200	13 421	14 164	14 912	15 664	16 420	17 181	17 945	18 713	19 484	20 259	21 036
300	21 036	21 817	22 600	23 386	24 174	24 964	25 757	26 552	27 348	28 146	28 946
400	28 946	29 747	30 550	31 354	32 159	32 965	33 772	34 579	35 387	36 196	37 005
500	37 005	37 815	38 624	39 434	40 243	41 053	41 862	42 671	43 479	44 286	45 093
600	45 093	45 900	46 705	47 509	48 313	49 116	49 917	50 718	51 517	52 315	53 112
700	53 112	53 908	54 703	55 497	56 289	57 080	57 870	58 659	59 446	60 232	61 017
800	61 017	61 801	62 583	63 364	64 144	64 922	65 698	66 473	67 246	68 017	68 787
900	68 787	69 554	70 319	71 082	71 844	72 603	73 360	74 115	74 869	75 621	76 373

Table B.9: Standard reference data for **type J** thermocouples. Data are values of thermocouple emf, in μV, for various tip temperatures, assuming a cold junction at 0°C.

°C	0	10	20	30	40	50	60	70	80	90	100
−200	−7 890	−8 095									
−100	−4 633	−5 037	−5 426	−5 801	−6 159	−6 500	−6 821	−7 123	−7 403	−7 659	−7 890
0	0	−501	−995	−1 482	−1 961	−2 431	−2 893	−3 344	−3 786	−4 215	−4 633
0	0	507	1 019	1 537	2 059	2 585	3 116	3 650	4 187	4 726	5 269
100	5 269	5 814	6 360	6 909	7 459	8 010	8 562	9 115	9 669	10 224	10 779
200	10 779	11 334	11 889	12 445	13 000	13 555	14 110	14 665	15 219	15 773	16 327
300	16 327	16 881	17 434	17 986	18 538	19 090	19 642	20 194	20 745	21 297	21 848
400	21 848	22 400	22 952	23 504	24 057	24 610	25 164	25 720	26 276	26 834	27 393
500	27 393	27 953	28 516	29 080	29 647	30 216	30 788	31 362	31 939	32 519	33 102
600	33 102	33 689	34 279	34 873	35 470	36 071	36 675	37 284	37 896	38 512	39 132
700	39 132	39 755	40 382	41 012	41 645	42 281	42 919	43 559	44 203	44 848	45 494
800	45 494	46 141	46 786	47 431	48 074	48 715	49 353	49 989	50 622	51 251	51 877
900	51 877	52 500	53 119	53 735	54 347	54 956	55 561	56 164	56 763	57 360	57 953

REFERENCE FUNCTIONS

Table B.10: Standard reference data for **type K** thermocouples. Data are values of thermocouple emf, in μV, for various tip temperatures, assuming a cold junction at 0°C.

°C	0	10	20	30	40	50	60	70	80	90	100
−200	−5 891		−6 158	−6 262	−6 344	−6 404	−6 441	−6 458			
−100	−3 554	−6 035	−4 138	−4 411	−4 669	−4 913	−5 141	−5 354	−5 550	−5 730	−5 891
0	0	−3 852	−778	−1 156	−1 527	−1 889	−2 243	−2 587	−2 920	−3 243	−3 554
		−392									
0	0	397	798	1 203	1 612	2 023	2 436	2 851	3 267	3 682	4 096
100	4 096	4 509	4 920	5 328	5 735	6 138	6 540	6 941	7 340	7 739	8 138
200	8 138	8 539	8 940	9 343	9 747	10 153	10 561	10 971	11 382	11 795	12 209
300	12 209	12 624	13 040	13 457	13 874	14 293	14 713	15 133	15 554	15 975	16 397
400	16 397	16 820	17 243	17 667	18 091	18 516	18 941	19 366	19 792	20 218	20 644
500	20 644	21 071	21 497	21 924	22 350	22 776	23 203	23 629	24 055	24 480	24 905
600	24 905	25 330	25 755	26 179	26 602	27 025	27 447	27 869	28 289	28 710	29 129
700	29 129	29 548	29 965	30 382	30 798	31 213	31 628	32 041	32 453	32 865	33 275
800	33 275	33 685	34 093	34 501	34 908	35 313	35 718	36 121	36 524	36 925	37 326
900	37 326	37 725	38 124	38 522	38 918	39 314	39 708	40 101	40 494	40 885	41 276
1000	41 276	41 665	42 053	42 440	42 826	43 211	43 595	43 978	44 359	44 740	45 119
1100	45 119	45 497	45 873	46 249	46 623	46 995	47 367	47 737	48 105	48 473	48 838
1200	48 838	49 202	49 565	49 926	50 286	50 644	51 000	51 355	51 708	52 060	52 410
1300	52 410	52 759	53 106	53 451	53 795	54 138	54 479	54 819			

Table B.11: Standard reference data for **type N** thermocouples. Data are values of thermocouple emf, in μV, for various tip temperatures, assuming a cold junction at 0°C.

°C	0	10	20	30	40	50	60	70	80	90	100
−200	−3 990	−4 083	−4 162	−4 226	−4 277	−4 313	−4 336	−4 345			
−100	−2 407	−2 612	−2 808	−2 994	−3 171	−3 336	−3 491	−3 634	−3 766	−3 884	−3 990
0	0	−260	−518	−772	−1 023	−1 269	−1 509	−1 744	−1 972	−2 193	−2 407
0	0	261	525	793	1 065	1 340	1 619	1 902	2 189	2 480	2 774
100	2 774	3 072	3 374	3 680	3 989	4 302	4 618	4 937	5 259	5 585	5 913
200	5 913	6 245	6 579	6 916	7 255	7 597	7 941	8 288	8 637	8 988	9 341
300	9 341	9 696	10 054	10 413	10 774	11 136	11 501	11 867	12 234	12 603	12 974
400	12 974	13 346	13 719	14 094	14 469	14 846	15 225	15 604	15 984	16 366	16 748
500	16 748	17 131	17 515	17 900	18 286	18 672	19 059	19 447	19 835	20 224	20 613
600	20 613	21 003	21 393	21 784	22 175	22 566	22 958	23 350	23 742	24 134	24 527
700	24 527	24 919	25 312	25 705	26 098	26 491	26 883	27 276	27 669	28 062	28 455
800	28 455	28 847	29 239	29 632	30 024	30 416	30 807	31 199	31 590	31 981	32 371
900	32 371	32 761	33 151	33 541	33 930	34 319	34 707	35 095	35 482	35 869	36 256
1000	36 256	36 641	37 027	37 411	37 795	38 179	38 562	38 944	39 326	39 706	40 087
1100	40 087	40 466	40 845	41 223	41 600	41 976	42 352	42 727	42 101	43 474	43 846
1200	43 846	44 218	44 588	44 958	45 326	45 694	46 060	46 425	46 789	47 152	47 513

REFERENCE FUNCTIONS

Table B.12: Standard reference data for **type R** thermocouples. Data are values of thermocouple emf, in μV, for various tip temperatures, assuming a cold junction at 0°C.

°C	0	10	20	30	40	50	60	70	80	90	100
0	0	54	111	171	232	296	363	431	501	573	647
100	647	723	800	879	959	1041	1124	1208	1294	1381	1469
200	1469	1558	1648	1739	1831	1923	2017	2112	2207	2304	2401
300	2401	2498	2597	2696	2796	2896	2997	3099	3201	3304	3408
400	3408	3512	3616	3721	3827	3933	4040	4147	4255	4363	4471
500	4471	4580	4690	4800	4910	5021	5133	5245	5357	5470	5583
600	5583	5697	5812	5926	6041	6157	6273	6390	6507	6625	6743
700	6743	6861	6980	7100	7220	7340	7461	7583	7705	7827	7950
800	7950	8073	8197	8321	8446	8571	8697	8823	8950	9077	9205
900	9205	9333	9461	9590	9720	9850	9980	10111	10242	10374	10506
1000	10506	10638	10771	10905	11039	11173	11307	11442	11578	11714	11850
1100	11850	11986	12123	12260	12397	12535	12673	12812	12950	13089	13228
1200	13228	13367	13507	13646	13786	13926	14066	14207	14347	14488	14629
1300	14629	14770	14911	15052	15193	15334	15475	15616	15758	15899	16040
1400	16040	16181	16323	16464	16605	16746	16887	17028	17169	17310	17451
1500	17451	17591	17732	17872	18012	18152	18292	18431	18571	18710	18849
1600	18849	18988	19126	19264	19402	19540	19677	19814	19951	20087	20222
1700	20222	20356	20488	20620	20749	20877	21003				

Table B.13: Standard reference data for **type S** thermocouples. Data are values of thermocouple emf, in μV, for various tip temperatures, assuming a cold junction at 0°C.

°C	0	10	20	30	40	50	60	70	80	90	100
0	0	55	113	173	235	299	365	433	502	573	646
100	646	720	795	872	950	1029	1110	1191	1273	1357	1441
200	1441	1526	1612	1698	1786	1874	1962	2052	2141	2232	2323
300	2323	2415	2507	2599	2692	2786	2880	2974	3069	3164	3259
400	3259	3355	3451	3548	3645	3742	3840	3938	4036	4134	4233
500	4233	4332	4432	4532	4632	4732	4833	4934	5035	5137	5239
600	5239	5341	5443	5546	5649	5753	5857	5961	6065	6170	6275
700	6275	6381	6486	6593	6699	6806	6913	7020	7128	7236	7345
800	7345	7454	7563	7673	7783	7893	8003	8114	8226	8337	8449
900	8449	8562	8674	8787	8900	9014	9128	9242	9357	9472	9587
1000	9587	9703	9819	9935	10051	10168	10285	10403	10520	10638	10757
1100	10757	10875	10994	11113	11232	11351	11471	11590	11710	11830	11951
1200	11951	12071	12191	12312	12433	12554	12675	12796	12917	13038	13159
1300	13159	13280	13402	13523	13644	13766	13887	14009	14130	14251	14373
1400	14373	14494	14615	14736	14857	14978	15099	15220	15341	15461	15582
1500	15582	15702	15822	15942	16062	16182	16301	16420	16539	16658	16777
1600	16777	16895	17013	17131	17249	17366	17483	17600	17717	17832	17947
1700	17947	18061	18174	18285	18395	18503	18609				

Table B.14: Standard reference data for **type T** thermocouples. Data are values of thermocouple emf, in μV, for various tip temperatures, assuming a cold junction at 0°C.

°C	0	10	20	30	40	50	60	70	80	90	100
−200	−5 603	−5 753	−5 888	−6 007	−6 105	−6 180	−6 232	−6 258			
−100	−3 379	−3 657	−3 923	−4 177	−4 419	−4 648	−4 865	−5 070	−5 261	−5 439	−5 603
0	0	−383	−757	−1 121	−1 475	−1 819	−2 153	−2 476	−2 788	−3 089	−3 379
0	0	391	790	1 196	1 612	2 036	2 468	2 909	3 358	3 814	4 279
100	4 279	4 750	5 228	5 714	6 206	6 704	7 209	7 720	8 237	8 759	9 288
200	9 288	9 822	10 362	10 907	11 458	12 013	12 574	13 139	13 709	14 283	14 862
300	14 862	15 445	16 032	16 624	17 219	17 819	18 422	19 030	19 641	20 255	20 872
400	20 872										

Table B.15: Reverse coefficients for type T thermocouples, for converting emf in μV to temperature (ITS–90) in °C using the expression $T = \sum c_j V^j$, an inverse equivalent of the standard reference function to ~ 0.05°C.

Temperature Range	j	c_j	Temperature Range	j	c_j
-200 to 0°C	0	0	0 to 400°C	0	0
	1	2.595×10^{-2}		1	2.593×10^{-2}
	2	-2.132×10^{-7}		2	-7.603×10^{-7}
	3	7.902×10^{-10}		3	4.638×10^{-11}
	4	4.253×10^{-13}		4	-2.166×10^{-15}
	5	1.3305×10^{-16}		5	6.048×10^{-20}
	6	2.0241×10^{-20}		6	-7.29×10^{-25}
	7	1.2668×10^{-24}			

Table B.16: Reverse coefficients for type K thermocouples, for converting emf in μV to temperature (ITS–90) in °C using the expression $T = \sum c_j V^j$, an inverse equivalent of the standard reference function to ~ 0.05°C.

Temperature Range	j	c_j	Temperature Range	j	c_j
-200 to 0°C	0	0	0 to 500°C	0	0
	1	2.5174×10^{-2}		1	2.5084×10^{-2}
	2	-1.166×10^{-6}		2	7.86×10^{-8}
	3	-1.0833×10^{-9}		3	-2.5031×10^{-10}
	4	-8.9773×10^{-13}		4	8.3153×10^{-14}
	5	-3.73423×10^{-16}		5	$-1.228034 \times 10^{-17}$
	6	-8.66327×10^{-20}		6	9.80404×10^{-22}
	7	-1.04505×10^{-23}		7	-4.41303×10^{-26}
	8	-5.192×10^{-28}		8	1.05773×10^{-30}
				9	-1.05275×10^{-35}
500 to 1372°C	0	-1.318×10^{2}			
	1	4.8302×10^{-2}			
	2	-1.646×10^{-6}			
	3	5.4647×10^{-11}			
	4	-9.6507×10^{-16}			
	5	8.8022×10^{-21}			
	6	-3.1111×10^{-26}			

Bibliography

[1] T. J. Seebeck. Evidence of the thermal current of the combination Bi-Cu by its action on magnetic needle. *Royal Acad. Sciences, Berlin*, pages 265–373, 1822-23.

[2] B. S. Finn. Thermoelectricity. *Adv. Electron. Electron. Phys.*, 50:175–240, 1980.

[3] C. S. M. Pouillet. Recherches sur les hautes temperatures et sur plusieurs phenomenes qui en dependent. *Compt. Rend.*, 3:782–90, 1836.

[4] L. B. Hunt. The early history of the thermocouple. *Plat. Metals Rev.*, 8:23–28, 1964.

[5] P. A. Kinzie. *Thermocouple Temperature Measurement*. John Wiley and Sons, New York, 1973.

[6] R. J. Moffat. The gradient approach to thermocouple circuitry. In A. I. Dahl, editor, *Temperature: its Measurement and Control in Science and Industry, Vol. 3 Pt. 2*, pages 33–38. Reinhold, New York and London, 1962.

[7] R. E. Bentley. Understanding thermoelectricity. Aust. Inst. Phys. Temperature Measurement Symposium, University of NSW, Sydney, Aug 1966.

[8] A. W. Fenton. How do thermocouples work? *Nucl. Energy*, 19:61–63, 1980.

[9] R. E. Bentley. The distributed nature of emf in thermocouples and its consequences. *Aust. J. Instrum. Control*, 38:128–32, 1982.

[10] W. F. Roeser. Thermoelectric thermometry. *J. Appl. Phys.*, 11:388–407, 1940.

[11] ASTM STP 470A. Manual on the use of thermocouples in temperature measurement. Special Tech. Publ. 470A, Am. Soc. for Testing and Materials, Philadelphia, 1974.

[12] R. L. Powell, W. J. Hall, C. H. Hyink Jr., and L. L. Sparks. Thermocouple reference tables based on the IPTS–68. NBS Monogr. 125, Natl. Bur. Stand., USA, 1974.

[13] N. A. Burley, R. L. Powell, G. W. Burns, and M. G. Scroger. The Nicrosil versus Nisil thermocouple: Properties and thermoelectric reference data. NBS Monogr. 161, Natl. Bur. Stand., USA, 1978.

[14] G. W. C. Kaye and T. H. Laby (Ed.). *Tables of Physical and Chemical Constants*. Longman, London and New York, 1973.

[15] R. B. Roberts. The absolute scale of thermoelectricity II. *Phil. Mag. B*, 43(6):1125–35, 1981.

[16] N. Cusack and P. Kendall. The absolute scale of thermoelectric power at high temperatures. *Proc. Phys. Soc.*, 72:898–901, 1958.

[17] R. B. Roberts, F. Righini, and R. C. Compton. The absolute scale of thermoelectricity III. *Phil. Mag. B*, 52(6):1147–63, 1985.

[18] M. V. Vedernikov. The thermoelectric powers of transition metals at high temperatures. *Adv. Phys.*, 18:337–70, 1969.

[19] R. E. Bentley. The thermocouple: a modern view. *Metals Australas.*, 13:19–25, 1981.

[20] R. E. Bentley. Short-term instabilities in thermocouples containing nickel-based alloys. *High Temp.-High Pressures*, 15:599–611, 1983.

[21] G. W. Burns, M. G. Scroger, G. F. Strouse, M. C. Croarkin, and W. F. Guthrie. Temperature-electromotive force reference functions and tables for the letter-designated thermocouple types based on the ITS–90. NIST Monogr. 175, National Institute of Standards and Technology, USA, 1993.

[22] ASTM E230. Temperature-electromotive force (emf) tables for standardised thermocouples. Annual book ASTM standards: vol. 14.03, Am. Soc. for Testing and Materials, Philadelphia, 1996.

[23] R. E. Bedford, G. Bonnier, H. Maas, and F. Pavese. Recommended values of temperature on ITS–90 for a selected set of secondary reference points. *Metrologia*, 33:133–54, 1996.

[24] Y. S. Touloukian, R. K. Kirby, R. E. Taylor, and P. D. Desai. *Thermophysical Properties of Matter*, volume 12. IFI/Plenum, New York, 1975.

[25] A. Goldsmith, T. E. Waterman, and H. J. Hirschhorn. *Handbook of Thermophysical Properties of Solid Materials*, volume 1. Macmillan Co, New York, 1961.

[26] D. R. Lide (Ed.). *Handbook of Chemistry and Physics*. CRC Press, Boca Raton, 1992–93.

[27] J. P. Moore, R. S. Graves, M. B. Herskovitz, K. R. Carr, and R. A. Vandermeer. Nicrosil II and Nisil thermocouple alloys: physical properties and behaviour during thermal cycling to 1200 K. In P. G. Klemens and T. K. Chu, editors, *Proc. 14th Int. Conf. on Thermal Cond.*, pages 259–66. Plenum Press, New York and London, 1975.

[28] Hoskins. *Chromel-Alumel Thermocouple Alloys*. Hamburg, Michigan, 1961. Hoskins Manufacturing Co Catalogue M-61 C-A.

[29] Hoskins. *Nicrosil/Nisil Thermocouple Alloys*. Hamburg, Michigan, 1986. Hoskins Manufacturing Co Catalogue.

[30] E. H. McLaren and E. G. Murdock. The Pt/Au thermocouple. Report NRCC/27703, National Research Council Canada, Ottawa, 1987.

[31] R. E. Bentley. The use of elemental thermocouples in high-temperature precision thermometry. *Measurement* - in print, 1998.

[32] G. W. Burns, G. F. Strouse, B. M. Liuand, and B. W. Mangum. Gold versus platinum thermocouples: performance data and an ITS–90 based reference function. In J. F. Schooley, editor, *Temperature: its Measurement and Control in Science and Industry, Vol. 6*, pages 531–6. Am. Inst. Physics, New York, 1992.

[33] G. W. Burns and D. C. Ripple. Techniques for fabricating and annealing Pt/Pd thermocouples for accurate measurements in the range 0 to 1300°C. In *Proc. of TEMPMEKO 96: 6th International Symposium on Temperature and Thermal Measurements in Industry and Science*, pages 171–76, Torino, Italy, September 1996.

[34] A. S. Darling and G. L. Selman. Some effects of environment on the performance of noble metal thermocouples. In H. H. Plumb, editor, *Temperature: its Measurement and Control in Science and Industry, Vol. 4*, pages 1633–44. ISA, Pittsburgh, 1972.

[35] E. H. McLaren and E. G. Murdock. New considerations on the preparation, properties and limitations of the standard thermocouple for thermometry. In H. H. Plumb, editor, *Temperature: its Measurement and Control in Science and Industry, Vol. 4*, pages 1543–60. ISA, Pittsburgh, 1972.

[36] R. E. Bentley and T. L. Morgan. Thermoelectric effects of cold work in Pt 10%Rh and Pt 13%Rh versus Pt thermocouples. *Metrologia*, 20:61–66, 1984.

[37] R. E. Bentley and T. P. Jones. Inhomogeneities in type S thermocouples when used to 1064°C. *High Temp.-High Pressures*, 12:33–45, 1980.

[38] R. E. Bentley. Changes in Seebeck coefficient of Pt and Pt 10%Rh after use to 1700°C in high-purity polycrystalline alumina. *Int. J. Thermophys.*, 6:83–99, 1985.

[39] B. E. Walker, C. T. Ewing, and R. R. Miller. Thermoelectric instability of some noble metal thermocouples at high temperature. *Rev. Sci. Instrum.*, 33:1029–40, 1962.

[40] B. E. Walker, C. T. Ewing, and R. R. Miller. Study of the instability of noble metal thermocouples in vacuum. *Rev. Sci. Instrum.*, 36:601–06, 1965.

[41] G. E. Glawe and A. J. Szaniszlo. Long-term drift of some noble- and refractory-metal thermocouples at 1600 K in air, argon and vacuum. In H. H. Plumb, editor, *Temperature: its Measurement and Control in Science and Industry, Vol. 4*, pages 1645–62. ISA, Pittsburgh, 1972.

[42] G. L. Selman. On the stability of metal sheathed noble metal thermocouples. In H. H. Plumb, editor, *Temperature: its Measurement and Control in Science and Industry, Vol. 4*, pages 1833–40. ISA, Pittsburgh, 1972.

[43] R. L. Anderson, J. D. Lyons, T. G. Kollie, W. H. Christie, and R. Eby. Decalibration of sheathed thermocouples. In J. F. Schooley, editor, *Temperature: its Measurement and Control in Science and Industry, Vol. 5*, pages 977–1007. Am. Inst. Phys., New York, 1982.

[44] R. J. Freeman. Thermoelectric stability of platinum vs platinum-rhodium thermocouples. In A. I. Dahl, editor, *Temperature: its Measurement and Control in Science and Industry, Vol. 3 Pt. 2*, pages 201–20. Reinhold, New York and London, 1962.

[45] E. D. Zysk and A. R. Robertson. Newer thermocouple materials. In H. H. Plumb, editor, *Temperature: its Measurement and Control in Science and Industry, Vol. 4*, pages 1697–734. ISA, Pittsburgh, 1972.

[46] R. E. Bentley. New-generation temperature probes. *Mater. Australas.*, 19(3):10–13, 1987.

[47] A. I. Dahl. Stability of base-metal thermocouples in air from 800 to 2200 °F. *J. Res. Natl. Bur. Stds*, 24:205–24, 1940.

[48] D. L. McElroy. Progress report 1: thermocouple research report - Nov 1 1956 to Oct 31 1957. ORNL-2467, 1957.

[49] J. F. Potts Jr. and D. L. McElroy. Thermocouple research to 1000°C - final report Nov 1 1957 through June 30 1959. ORNL-2773, 1961.

[50] P. C. Hughes and N. A. Burley. Metallurgical factors affecting stability of nickel-base thermocouples. *J. Inst. Metals*, 91:373–76, 1962-3.

[51] N. A. Burley and R. G. Ackland. The stability of the thermo-emf temperature characteristics of nickel-base thermocouples. *J. Aust. Inst. Metals*, 12:23–31, 1967.

[52] V. A. Callcut. Aging of Chromel Alumel thermocouples. UKAEA report 1021(R/X), 1965.

[53] A. W. Fenton, R. Dacey, and E. J. Evans. Thermocouples: instabilities of Seebeck coefficient. UKAEA TRG report 1447(R), 1967.

[54] R. Nordheim and N. J. Grant. Resistivity anomalies in the nickel-chromium system as evidence of ordering reactions. *J. Inst. Metals*, 82:440–44, 1953.

[55] D. D. Pollock. *Thermocouples Theory and Practice*, pages 225–8. CRC Press, Boca Raton, Florida, 1991.

[56] E. W. Northover and J. A. Hitchcock. The effect of heating on the thermoelectric power of commercial thermocouple wires. *Instrum. Practice*, 22:606–11, 1968.

[57] N. A. Burley. Cyclic thermo-emf drift in nickel chromium thermocouple alloys attributable to short range order. DSL report 353, 1970.

[58] A. W. Fenton. Errors in thermoelectric thermometers. *Proc. IEE*, 116:1277–85, 1969.

[59] E. W. Northover and J. A. Hitchcock. A new high-stability nickel alloy thermocouple. *Instrum. Practice*, 25:529–31, 1971.

[60] F. Hall and E. F. McGuire. Hoskins Manufacturing Co. Private communication, 1987.

[61] C. D. Starr and T. P. Wang. A new stable nickel-base thermocouple. *J. Testing and Eval.*, 42:42–56, 1976.

BIBLIOGRAPHY

[62] Materials Research Laboratories, DSL Annual Report, p. 71. Melbourne, 1967–68.

[63] N. A. Burley. Nicrosil and Nisil: highly stable nickel-base alloys for thermocouples. In H. H. Plumb, editor, *Temperature: its Measurement and Control in Science and Industry, Vol. 4*, pages 1677–95. ISA, Pittsburgh, 1972.

[64] N. A. Burley and T. P. Jones. Practical performance of Nicrosil Nisil thermocouples. *Inst. Phys. Conf. Ser. No. 26*, pages 172–80, 1975.

[65] T. P. Wang and D. Bediones. 10,000 h stability test of types K, N and a Ni-Mo/Ni-Co thermocouple in air and short-term tests in reducing atmospheres. In J. F. Schooley, editor, *Temperature: its Measurement and Control in Science and Industry, Vol. 6*, pages 595–600. Am. Inst. Physics, New York, 1992.

[66] G. W. Burns. The Nicrosil versus Nisil thermocouple: recent developments and present status. In J. F. Schooley, editor, *Temperature: its Measurement and Control in Science and Industry, Vol. 5*, pages 1121–27. Am. Inst. Phys., New York, 1982.

[67] R. E. Bentley. Thermoelectric hysteresis in Nicrosil and Nisil. *J. Phys. E.*, 20:1368–73, 1987.

[68] T. P. Wang, C. D. Starr, and N. Brown. Thermoelectric characteristics of binary alloys of nickel. *Acta Metall.*, 14:649–57, 1966.

[69] C. D. Starr and T. P. Wang. Effect of oxidation on the stability of thermocouples. *Proc. Am. Soc. Testing and Materials*, 63:1185–94, 1963.

[70] R. E. Bentley. The new Nicrosil-sheathed type N MIMS thermocouple: an assessment of the first production batch. *Mater. Australas.*, 18(6):16–18, 1986.

[71] R. E. Bentley. Irreversible thermoelectric changes in type K and type N thermocouple alloys within Nicrosil-sheathed MIMS cable. *J. Phys. D.*, 22:1908–15, 1989.

[72] N. A. Burley, J. L. Cocking, G. W. Burns, and M. G. Scroger. The Nicrosil versus Nisil thermocouple: the influence of magnesium on the thermoelectric stability and oxidation resistance of the alloys. In J. F. Schooley, editor, *Temperature: its Measurement and Control in Science and Industry, Vol. 5*, pages 1129–45. Am. Inst. Phys., New York, 1982.

[73] T. P. Wang and C. D. Starr. Oxidation resistance and stability of Nicrosil-Nisil in air and in reducing atmospheres. In J. F. Schooley, editor, *Temperature: its Measurement and Control in Science and Industry, Vol. 5*, pages 1147–57. Am. Inst. Phys., New York, 1982.

[74] C. Brookes, T. R. D. Chandler, and B. Chu. Nicrosil-Nisil: a new high stability thermocouple for the industrial user. *Meas. Control*, 18:245–48, 1985.

[75] R. E. Bentley and G. F. Russell. Nicrosil-sheathed mineral-insulated type N thermocouple probes for short-term variable-immersion applications to 1100°C. *Sensors and Actuators*, 16:89–100, 1989.

[76] R. E. Bentley. The case for MIMS thermocouples: a comparison with the bare-wire alternatives. *Process & Contr. Eng.*, 39(3):36–46, 1986.

[77] R. M. Hess. X-ray assessment of MIMS thermocouples. Commercial in confidence, May 1984.

[78] H. L. Daneman. MIMS thermocouple stability. *Measurements & Control*, 25(1):93–95, 1991.

[79] W. H. Christie, R. E. Eby, and R. L. Anderson. Ion microprobe investigation of decalibrated Chromel versus Alumel thermocouples. *Appl. Surface Science*, 3:329–47, 1979.

[80] R. E. Bentley and T. L. Morgan. Ni-based thermocouples in the mineral-insulated metal-sheathed format: thermoelectric instabilities to 1100°C. *J. Phys. E.*, 19:262–68, 1986.

[81] F. Andersen and N. A. Burley. Thermoelectric instability of some metal-sheathed mineral-insulated standard thermocouples related to type, sheath alloy and environment of exposure. In *Measurement and Progress*, pages 994–9. Proc. XIIth IMEKO World Congress, Beijing, Sept. 1991.

[82] R. E. Bentley. An advance in thermocouple design and its significance to heat treatment. *Mater. Australas.*, 21(7):10–15, 1989.

[83] R. E. Bentley. An enhanced temperature probe that complies with the type K thermocouple specification. In *Proc. Australas. Instrum. & Meas. Conf.*, pages 96–101, Adelaide, November 1989. Institution of Engineers Aust.

[84] R. E. Bentley. Thermoelectric behaviour of Ni-based ID-MIMS thermocouples using the Nicrosil-plus sheathing alloy. In J. F. Schooley, editor, *Temperature: its Measurement and Control in Science and Industry, Vol. 6*, pages 585–90. Am. Inst. Physics, New York, 1992.

[85] N. Fawcett and B. Wilson. High-temperature stability and endurance tests on bare-wire and mineral-insulated type N thermocouples. *High Temp.-High Pressures*, 22:439–48, 1990.

[86] R. E. Bentley. Optimising the thermoelectric stability of ID-MIMS type K thermocouples by adjusting the levels of Mn and Al. In J. F. Schooley, editor, *Temperature: its Measurement and Control in Science and Industry, Vol. 6*, pages 591–4. Am. Inst. Physics, New York, 1992.

[87] R. E. Bentley. Thermoelectric hysteresis in nickel-based thermocouple alloys. *J. Phys. D.*, 22:1902–7, 1989.

[88] R. E. Bentley. Recent developments in thermocouple design. *Aust. J. Instrum. Control*, 5(1):4–8, 1990.

[89] R. E. Bentley. Long-term drift in mineral-insulated Nicrosil-sheathed type K thermocouples. *Sensors and Actuators A*, 24:21–26, 1990.

[90] K. G. Parker. Thermal stability and oxidation resistance of Nicrobell type N MIMS thermocouples and BPTX type N thermocouples. BICC Thermoheat, report DS/90/1. 1990.

[91] R. L. Rusby, D. F. Carter, and A. Beswick. An evaluation of sheathed Nicrosil/Nisil thermocouples up to 1300°C. *Mater. High Temp.*, 10:193–200, 1992.

[92] N. A. Burley. Advanced integrally-sheathed type N thermocouple of ultra-high thermoelectric stability. In *Australas. Instrum. & Meas. Conf.*, pages 90–5. Institution of Engineers Aust., Adelaide, Nov. 1989.

[93] N. A. Burley. N-Clad-N a novel integrally-sheathed thermocouple: optimum design rationale for ultra-high thermoelectric stability. In J. F. Schooley, editor, *Temperature: its Measurement and Control in Science and Industry, Vol. 6*, pages 579–84. Am. Inst. Physics, New York, 1992.

[94] Hoskins. *Hoskins 2300 MI Cable.* Hamburg, Michigan, 1995. Hoskins Manufacturing Co Catalogue.

[95] K. K. Montgomery. Type N versus type K thermocouple comparison in a brick kiln. In J. F. Schooley, editor, *Temperature: its Measurement and Control in Science and Industry, Vol. 6*, pages 601–5. Am. Inst. Physics, New York, 1992.

[96] N. A. Burley, R. M. Hess, C. F. Howie, and J. A. Coleman. The Nicrosil versus Nisil thermocouple: a critical comparison with the ANSI standard letter-designated base-metal thermocouples. In J. F. Schooley, editor, *Temperature: its Measurement and Control in Science and Industry, Vol. 5*, pages 1159–66. Am. Inst. Phys., New York, 1982.

[97] J. G. Hust, R. L. Powell, and L. L. Sparks. Methods for cryogenic thermocouple thermometry. In H. H. Plumb, editor, *Temperature: its Measurement and Control in Science and Industry, Vol. 4*, pages 1525–35. ISA, Pittsburgh, 1972.

[98] L. M. Besley. Cryogenic thermometry. In R. E. Bentley, editor, *Temperature and Humidity Measurements*, volume 1 of *Handbook of Temperature Measurement*, chapter 3. Springer-Verlag, Singapore, 1998.

[99] ASTM E988. Temperature-electromotive force (emf) tables for tungsten-rhenium thermocouples. Annual book of ASTM standards: vol. 14.03, Am. Soc. for Testing and Materials, Philadelphia, 1988.

[100] J. C. Lachman and J. A. McGurty. The use of refractory metals for ultra high-temperature thermocouples. In A. I. Dahl, editor, *Temperature: its Measurement and Control in Science and Industry, Vol. 3 Pt. 2*, pages 177–88. Reinhold, New York and London, 1962.

[101] Hoskins. *Tungsten-Rhenium Thermocouple Alloys.* Hamburg, Michigan, 1974. Hoskins Manufacturing Co Catalogue.

[102] C. D. Starr and T. P. Wang. Thermocouples and extension wires. In H. H. Plumb, editor, *Temperature: its Measurement and Control in Science and Industry, Vol. 4*, pages 1781–87. ISA, Pittsburgh, 1972.

[103] G. Rosengarten. Physics of temperature measurement. In R. E. Bentley, editor, *Temperature and Humidity Measurements*, volume 1 of *Handbook of Temperature Measurement*, chapter 2. Springer-Verlag, Singapore, 1998.

[104] J. P. Holman. *Heat Transfer.* McGraw-Hill Book Co., Singapore, 1989.

[105] M. Jakob. *Heat Transfer, Vol. II.* John Wiley and Sons, New York, 1957.

[106] R. B. Barefoot and A. H. Zerban. *Combustion*, 21(4):53, 1949.

[107] T. Lamb and R. Barber. Suction pyrometers in theory and practice. *J. Iron Steel Inst.*, 184:269–73, 1956.

[108] R. E. Bentley. More efficient heat treatment through improved thermocouple application. *Heat Treat. Metals*, 17:46–50, 1990.

[109] G. Sandars and M. J. Ballico. Traceable measurements. In R. E. Bentley, editor, *Temperature and Humidity Measurements*, volume 1 of *Handbook of Temperature Measurement*, chapter 1. Springer-Verlag, Singapore, 1998.

[110] R. E. Bentley. An approach to thermocouple calibration based on the distributed nature of thermo-emf. In P. L. Hewitt, editor, *Modern Techniques in Metrology*, pages 228–44. World Scientific, Singapore, 1984.

[111] E. H. McLaren and E. G. Murdock. The properties of Pt/PtRh thermocouples for thermometry in the range 0 to 1000°C: effect of heat treatment. Report NRCC/17408, National Research Council Canada, Ottawa, 1979.

[112] G. W. Burns, G. F. Strouse, B. W. Mangum, M. C. Croarkin, W. F. Guthrie, P. Marcarino, M. Battuello, H. K. Lee, J. C. Kim, K. S. Gam, C. Rhee, M. Chattle, M. Arai, H. Sakurai, A. I. Pokhodun, N. P. Moiseeva, S. A. Perevalova, M. J. de Groot, J. Zhang, K. Fan, and S. Wu. New reference functions for platinum-10% rhodium versus platinum (type S) thermocouples based on the ITS-90. In J. F. Schooley, editor, *Temperature: its Measurement and Control in Science and Industry, Vol. 6*, pages 537–46. Am. Inst. Physics, New York, 1992.

[113] R. E. Bentley. Thermocouple standards for the NML automatic calibration facility. *CSIRO NML Tech. Memorandum 145*, 1997.

[114] R. E. Bentley. Variability of deviation functions and ease of interpolation for Pt-based thermocouples: assessed with the NML automatic calibration facility. *Metrologia*, 35(1):41–47, 1998.

[115] R. E. Bentley. Thermoelectric scanning and automatic calibration of thermocouples at NML. *CSIRO Tech. Memorandum 149*, 1997.

[116] J. J. Connolly. Industrial resistance thermometers. In R. E. Bentley, editor, *Resistance & Liquid-in-Glass Thermometry*, volume 2 of *Handbook of Temperature Measurement*, chapter 2. Springer-Verlag, Singapore, 1998.

[117] E. C. Horrigan. Calibration enclosures. In R. E. Bentley, editor, *Resistance & Liquid-in-Glass Thermometry*, volume 2 of *Handbook of Temperature Measurement*, chapter 8. Springer-Verlag, Singapore, 1998.

[118] R. E. Bentley and S. R. Meszaros. A laboratory diagnostic furnace for scanning and calibrating thermocouples. *Aust. J. Instrum. Control*, 4(4):4–9, 1989.

[119] R. E. Bentley. The thermocouple in quality control. In *Proc. 35 Ann. Conf. Australas. Inst. Metals*, pages 152–55, Sydney, May 1982.

[120] *ISO Guide to the Expression of Uncertainty in Measurement*. International Organization for Standardization, Geneva, 1993.

[121] AS 2706. Numerical values - rounding and interpretation of limiting values. Australian standard 2706, Standards Australia, Sydney, 1984.

[122] BS 1957. Presentation of numerical values (fineness of expression; rounding of numbers). British standard 1957, British Standards Institution, London, 1953.

[123] ASTM E29-67. Calibration of thermocouples by comparison techniques. Annual book ASTM standards: vol. 3.01, Am. Soc. for Testing and Materials, Philadelphia, 1988.

[124] W. J. Dixon and F. J. Massey. *Introduction to Statistical Analysis.* McGraw-Hill Book Co., New York, 1957.

[125] R. H. Gassner. Heat treating aluminum alloy aircraft parts. *Metal Progr.*, 67:75–79, 1955.

[126] R. E. Bentley and T. P. Jones. A method for the determination of temperature variations within heat treatment installations with emphasis on the treatment of uncertainties. *Aust. J. Metals*, 12:302–09, 1967.

[127] O. Lutherer and R. J. Reed. Method for improving temperature uniformity in furnaces. *Metal Progr.*, 65:113, 1954.

[128] AS 2853. Enclosures, temperature-controlled: Performance testing and grading. Australian Standard 2853, Standards Australia, Sydney, 1986.

[129] R. E. Bentley. The NSL approach to testing heated enclosures. Proceedings of NATA sponsored meeting on Heated Enclosures, National Science Centre, Melbourne, 10 April 1974.

[130] H. B. Callen. The application of Onsager's reciprocal relations to thermoelectric and galvano-magnetic effects. *Phys. Rev.*, 73:1349–58, 1948.

[131] E. F. Obert. *Concepts of Thermodynamics.* McGraw-Hill, New York, 1960.

[132] R. Stratton. On the elementary theory of thermoelectric phenomena. *Brit. J. Appl. Phys.*, 8:315–21, 1957.

[133] R. G. Chambers. Thermoelectric effects and contact potentials. *Phys. Educ.*, 12:374–80, 1977.

[134] R. B. Roberts. The absolute scale of thermoelectricity. *Phil. Mag.*, 36(1):91–107, 1977.

[135] C. W. Lee, J. R. Kuhn, C. L. Foiles, and J. Bass. Thermopower and magnetothermopower of PtCo. In F. J. Blatt and P. A. Schroeder, editors, *Thermoelectricity in Metallic Conductors*, pages 307–13. Plenum Press, New York and London, 1978.

[136] J. Nyström. Thomson coefficients for copper at high temperatures. *Ark. Mat. Astr. Fys.*, 34A:No. 27, 1948.

[137] C. Kittel. *Introduction to Solid State Physics.* John Wiley and Sons, New York, 1956.

[138] H. Preston-Thomas. The International Temperature Scale of 1990 (ITS–90). *Metrologia*, 27:3–10, 1990.

[139] The International Temperature Scale of 1990. In R. E. Bentley, editor, *Resistance & Liquid-in-Glass Thermometry*, volume 2 of *Handbook of Temperature Measurement*, chapter 9. Springer-Verlag, Singapore, 1998.